Soil Mechanics Through Project-Bas
Learning

Soil Mechanics Through Project-Based Learning

Ivan Gratchev, Dong-Sheng Jeng
and Erwin Oh

Griffith School of Engineering Gold Coast, Southport, QLD, Australia

CRC Press
Taylor & Francis Group
Boca Raton   London   New York   Leiden

CRC Press is an imprint of the
Taylor & Francis Group, an **informa** business

A BALKEMA BOOK

CRC Press/Balkema is an imprint of the Taylor & Francis Group, an informa business

© 2019 Taylor & Francis Group, London, UK

Typeset by Apex CoVantage, LLC

Library of Congress Cataloging-in-Publication Data
Names: Gratchev, Ivan, author. | Oh, Erwin, author. | Jeng, Dong-Sheng, author.
Title: Soil mechanics through project-based learning / Ivan Gratchev, Erwin Oh & Dong-Sheng Jeng, Griffith School of Engineering Gold Coast, Southport, QLD, Australia.
Description: London ; Boca Raton : CRC Press/Balkema is an imprint of the Taylor & Francis Group, an Informa Business, [2019] | Includes bibliographical references and index.
Identifiers: LCCN 2018039215 (print) | LCCN 2018039939 (ebook) | ISBN 9780429507786 (ebook) | ISBN 9781138500075 (pbk) | ISBN 9781138605732 (hbk)
Subjects: LCSH: Soil mechanics—Textbooks. | Active learning—Textbooks.
Classification: LCC TA710 (ebook) | LCC TA710 .G625 2019 (print) | DDC 624.1/5136—dc23
LC record available at https://lccn.loc.gov/2018039215

Published by: CRC Press/Balkema
                Schipholweg 107c, 2316 XC Leiden, The Netherlands
                e-mail: Pub.NL@taylorandfrancis.com
                www.crcpress.com – www.taylorandfrancis.com

ISBN: 978-1-138-50007-5 (Pbk)
ISBN: 978-1-138-60573-2 (Hbk)
ISBN: 978-0-429-50778-6 (eBook)

Contents

Preface

This book is written for students who would like to learn the fundamentals and practical aspects of soil mechanics using a more hands-on approach than traditional textbooks. There are several textbooks on soil mechanics on the market; however, most of these texts were written many years ago using the traditional format of presentation. Students would then be expected to rote learn and then regurgitate this information accordingly. However, our experience as teachers suggests that this traditional format where the instructor provides students with theoretical knowledge through a series of lectures and abstract textbook problems is not sufficient to prepare students to tackle real geotechnical challenges. There is a real disjunct between what students are taught in universities and what they are expected to do in practice as engineering practitioners. It is not an uncommon situation that students are able to derive complex equations in the classroom, but struggle in the workforce when faced with practical engineering challenges.

This book employs a more engaging project-based approach to learning, which partially simulates what practitioners do in real life. The project-based method, which has proven to be a valid alternative to the traditional one, not only provides students with opportunities to better understand the fundamentals of soil mechanics, but also allows them to gain experience that is more practical and learn how to apply theory to practice. Our teaching experience indicates that working on a practical project makes the learning process more relevant and engaging. This book will appeal to the new generations of students who would like to have a better idea of what to expect in their employment future. In this book, readers are presented with a real-world challenge (in the form of a project-based assignment) similar to that which they would encounter in engineering practice and they need to work-out solutions using the relevant theoretical concepts that are briefly summarized in the book chapters. To complete this project-based assignment, readers are required to undertake a series of major geotechnical tasks including: a) interpretation of field and laboratory data, b) analysis of soil conditions, c) identification of geotechnical problems at a construction site and d) assessment of their effect on construction.

This book covers all significant topics in soil mechanics and slope stability analysis. Each section is followed by several review questions that will reinforce the reader's knowledge and make the learning process more engaging. A few typical problems are discussed at the end of chapters to help the reader develop problem-solving skills. Once the reader has sufficient knowledge of soil properties and mechanics, they will be able to undertake a project-based assignment to scaffold their learning. The assignment is based on real field and laboratory data including boreholes and test results so that the reader can experience what geotechnical engineering practice is like, identify with it personally and integrate it into their own knowledge base. In addition, some problems will include open-ended questions, which will encourage the reader to exercise their judgment and develop practical skills. To foster the learning process, solutions to all questions will be provided and discussed.

We are grateful to all students of Soil Mechanics and Geotechnical engineering courses at Griffith University for their constructive feedback in the past several years. We are also grateful to Professor Arumugam Balasubramaniam for his continuous support and encouragement.

Conversion factors

Length	Mass and weight	Area	Volume	Unit weight	Stress
1 in = 2.54 cm 1 ft = 30.5 cm	1 lb = 454 g 1 lb = 4.46 N 1 lb = 0.4536 kgf	1 in^2 = 6.45 cm^2 1 ft^2 = 0.0929 m^2	1 ml = 1 cm^3 1 l = 1000 cm^3 1 ft^3 = 0.0283 m^3 1 in^3 = 16.4 cm^3	1 lb/ft^3 = 0.157 kN/m^3	1 lb/in^2 = 6.895 kN/m^2 1 lb/ft^2 = 47.88 N/m^2
1 m = 39.37 in 1 m = 3.281 ft	1 N = 0.2248 lb 1 metric ton = 2204.6 lb 1 kgf = 2.2046 lb	1 m^2 = 10.764 ft^2 1 cm^2 = 0.155 in^2	1 m^3 = 35.32 ft^3 1 cm^3 = 0.061023 in^3	1 kN/m^3 = 6.361 lb/ft^3	1 kN/m^2 = 20.885 lb/ft^2 1 kN/m^2 = 0.145 lb/in^2

Chapter 1

Introduction

Innovative approach used in this book. This book is written for students and those who would like to learn the fundamentals and practical applications of soil mechanics without having to go through numerous pages of "boring" theory and equations. It is intended to show the reader how we use soil mechanics in practice and why it is important to know soil behavior. Unlike existing books on soil mechanics, this one uses a different, project-based approach where the reader is introduced to a real geotechnical project, which was conducted to select the most appropriate place for canal construction. This project will underscore the need for the theory presented in this book. The reader will learn how to interpret and analyze the data from field and laboratory investigation as well as how to use it to solve various geotechnical problems. Every step of this project is related to a certain aspect of soil mechanics and as the reader acquires more knowledge in the process, they will be able to see the practical importance of what they have learnt and thus better understand it. Our teaching experience (Gratchev and Jeng, 2018) indicates that the understanding of basic soil properties and soil behavior is much more important than remembering how to derive theoretical equations or memorizing numbers. For this reason, we pay more attention to explaining important concepts and discussing step-by-step solutions to common geotechnical problems.

Book organization. The book is organized in such a way that each chapter first explains its relevance to the project and then briefly introduces key theoretical concepts necessary to complete a certain part of the project. Chapter 1 provides the project description and data from field investigation and laboratory testing. The following chapters deal with soil origin and exploration (Chapter 2), basic soil properties (Chapters 3–4), fundamental concepts of soil compaction (Chapter 5) and stresses in soil mass (Chapter 6) as well as more advanced geotechnical applications such as flow nets (Chapter 7), soil deformation and consolidation (Chapters 8–10) and shear strength and slope stability analysis (Chapter 11).

Material for self-practice. Each chapter provides practical problems that the reader should use for more practice. We suggest trying to solve each problem first before referring to the step-by-step solution provided afterwards. Even though it may be difficult to work it out all the way to the final answer, spending time on each problem will improve the reader's understanding of the relevant material and help to develop problem-solving skills. In addition, to reinforce the knowledge of soil behavior and review the key concepts, the reader can take a quiz at the end of each chapter. To make this book more interactive, we have included several questions (and answers) that we commonly receive from our students.

1.1 Project description

In 2009, a residential canal was proposed at the Pimpama Riverside Development and geotechnical investigation was performed to assess the geotechnical settings of the area. The results of this investigation include a map (Fig. 1.1), a series of boreholes logs (Figs. 1.2–1.4) and laboratory reports (Tables 1.1–1.3). The main objective of this investigation was to determine properties of soil and estimate potential natural hazard and geotechnical issues that might occur during construction. Previous engineering developments in that area revealed some issues related to widely spread very soft clay, which was not appropriate foundation for most of engineering structures. To address this problem, it was considered to consolidate this soft material using a pre-load method (this will be discussed in detail in Chapter 10).

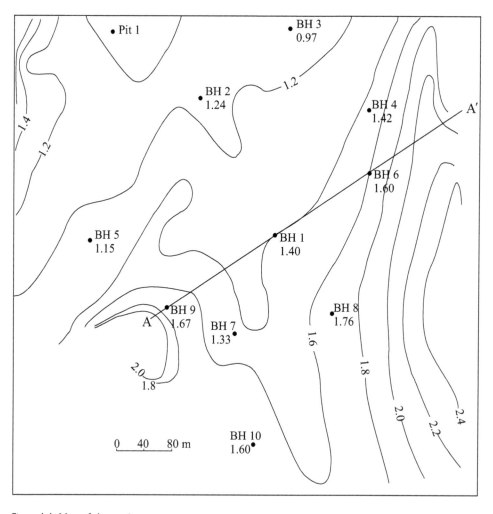

Figure 1.1 Map of the project area.

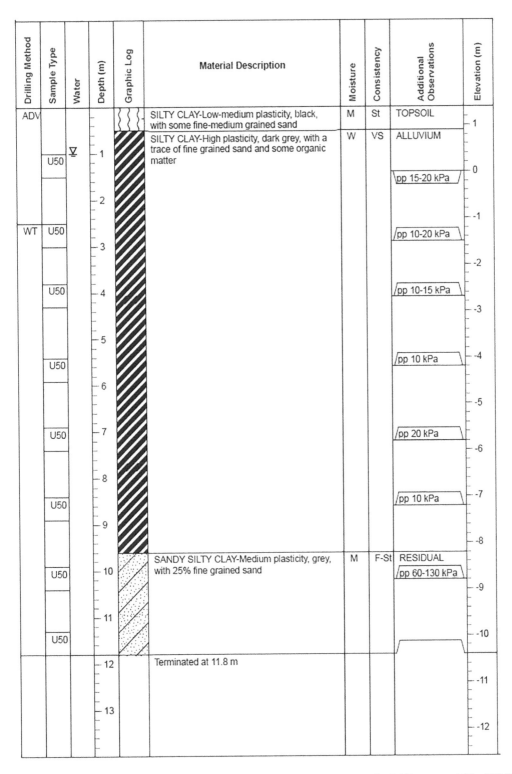

Figure 1.2 Borehole BH1, Elevation – 1.4 m. Borehole log legend: Drilling method: AD – auger drilling, V- "V" bit, W-washbore, T-TC bit. Sample type: U50-undisturbed sample 50 mm diameter. Moisture: M-moist, W-wet. Consistency: VS-very soft, F-firm, St-stiff. Additional observations: pp-pocket penetrometer.

Drilling Method	Sample Type	Water	Depth (m)	Graphic Log	Material Description	Moisture	Consistency	Additional Observations	Elevation (m)
WT					SILTY CLAY-Low-medium plasticity, dark, with some fine-medium grained sand	M	St	TOPSOIL	
					SILTY CLAY-High plasticity, dark grey and yellow, with a trace of fine grained sand and some shell fragments	W	VS	ALLUVIUM	1
			1						0
	U50		2						
								pp 10-15 kPa	-1
			3						
	U50								-2
			4					pp 15-20 kPa	
									-3
	U50		5					pp 20-25 kPa	
			6						-4
	U50							pp 20-35 kPa	-5
			7						
									-6
	U50		8		CLAY-Medium plasticity, grey, with some fine-medium grained sand	M-W	F	RESIDUAL	
								pp 75 kPa	
			9			M	St-V	pp 10 kPa	-7
	U50							pp 220 kPa	-8
			10		Terminated at 9.95 m				
			11						-9
			12						-10
			13						-11
									-12

Figure 1.3 Borehole BH6, Elevation – 1.6 m. Borehole log legend: Drilling method: W-washbore, T-TC bit. Sample type: U50-undisturbed sample 50 mm diameter. Moisture: M-moist, W-wet. Consistency: VS-very soft, F-firm, St-stiff. Additional observations: pp-pocket penetrometer.

Drilling Method	Sample Type	Water	Depth (m)	Graphic Log	Material Description	Moisture	Consistency	Additional Observations	Elevation (m)
WT					SILTY CLAY-Low-medium plasticity, dark brown, with some fine grained sand	M	F	TOPSOIL	1.5
			0.5		SILTY CLAY-High plasticity, dark grey, with a trace of fine grained sand and some shell fragments		S	ALLUVIUM	1
		⩔	1						0.5
			1.5			W			0
			2				VS		-0.5
	U50		2.5					pp 10 kPa	-1
			3						-1.5
			3.5						-2
	U50		4					pp 15-25 kPa	-2.5
			4.5						-3
			5						-3.5
	U50		5.5					pp 15-25 kPa	-4
			6						-4.5
			6.5						-5
	U50		7		CLAY-Medium plasticity, grey, with a trace of fine grained sand	M	VSt	RESIDUAL	-5.5
			7.5					pp 80 kPa	-6
	U50		8					pp 240-280 kPa	-6.5
			8.5		Terminated at 8.45 m				-7
			9						-7.5
			9.5						-8

Figure 1.4 Borehole BH9, Elevation – 1.67 m. Borehole log legend: Drilling method: W-washbore, T-TC bit. Sample type: U50-undisturbed sample 50 mm diameter. Moisture: M-moist, W-wet. Consistency: VS-very soft, S-soft, F-firm, St-stiff, VST-very stiff. Additional observations: pp-pocket penetrometer.

Table 1.1 Data from a dynamic cone penetration test. Location: near BH9.

Depth, cm	Number of blows, N_d	Depth, cm	Number of blows, N_d	Depth, cm	Number of blows, N_d
10	4	240	5	470	6
20	5	250	4	480	6
30	5	260	5	490	7
40	5	270	6	500	7
50	6	280	4	510	6
60	2	290	5	520	8
70	3	300	6	530	6
80	2	310	4	540	7
90	1	320	5	550	8
100	3	330	6	560	9
110	2	340	7	570	9
120	2	350	6	580	8
130	3	360	7	590	7
140	4	370	5	600	8
150	3	380	8	610	8
160	2	390	6	620	8
170	4	400	7	630	12
180	5	410	8	640	12
190	3	420	9	650	12
200	2	430	8	660	17
210	3	440	7	670	17
220	4	450	8	680	17
230	5	460	5	690	21

1.2 Field data

Field data for this project includes a site map (Fig. 1.1) that shows surface elevation contours and borehole location and borehole logs (Figs. 1.2–1.4). The borehole logs contain information about soil type, moisture and consistency, which is useful for overall assessment of site conditions. Table 1.1 presents data from a dynamic cone penetration test which was conducted near BH9.

1.3 Laboratory data

Laboratory data includes results of particle size distribution and Atterberg limits tests as well as information about soil *in-situ* density and moisture content. Three different soil types were tested: alluvial silty clay (Table 1.2), residual soil (Table 1.3) and sand from Pit 1 (Table 1.4).

Table 1.2 Particle size distribution and consistency limits test report (alluvium).

Client:	Home PTY LTD		Job No.	97638247
Project:	Riverside Development – Pimpama		Date Checked:	19-Jan-2009
Location:	BH1, Refer to Map		Checked by:	

Lab Reference No.	B15761	**Sample Identification:**	BH1: 2.60 – 3.00 m

Laboratory Specimen Description: Silty Clay (alluvium)
Sample size: Diameter 50 mm, height 100 mm

Particle Size Distribution AS1289 3.6.1

Sieve Size, mm	% Passing
9.5	
6.7	
4.75	100
2.36	99
1.18	98
0.600	96
0.425	94
0.300	91
0.150	88
0.075	83
0.020	80
0.006	76
0.002	73

Consistency Limits and Moisture Content

Test		Method	Result
Liquid Limit	%	AS1289 3.1.2	67
Plastic Limit	%	AS1289 3.2.1	29
Linear Shrinkage	%	AS1289 3.4.1	17
Moisture Content	%	AS1289 2.1.1	78.3
Wet Density	t/m³	AS1289 F4.1	1.47
Soil Particle Density	t/m³	AS1289 F4.1	2.53

Sample History:	Oven dried (100 deg.)
Preparation Method:	Dry sieved
Crumbling / Curling of linear shrinkage:	Nil
Linear shrinkage mould length:	250mm

Table 1.3 Particle size distribution and consistency limits test report (residual soil).

Client:	Home PTY LTD	Job No.	97638247
Project:	Riverside Development – Pimpama	Date Checked:	19-Jan-2009
Location:	BHI, Refer to Map	Checked by:	

| Lab Reference No. | B4577 | Sample Identification: | | BHI: 10.10 – 10.50 m |

Laboratory Specimen Description: Sandy Silty Clay (residual soil)
Sample size: Diameter 50 mm, height 100 mm

Particle Size Distribution AS1289 3.6.1

Sieve Size, mm	% Passing
19.0	100
13.2	96
9.5	91
6.7	87
4.75	81
2.36	75
1.18	70
0.600	63
0.425	58
0.300	52
0.150	43
0.075	35
0.020	24
0.006	17
0.002	12

Consistency Limits and Moisture Content

Test		Method	Result.
Liquid Limit	%	AS1289 3.1.2	29
Plastic Limit	%	AS1289 3.2.1	18
Linear Shrinkage	%	AS1289 3.4.1	4
Moisture Content	%	AS1289 2.1.1	19.1
Wet Density	t/m³	AS1289 F4.1	1.95
Soil Particle Density	t/m³	AS1289 F4.1	2.63
Sample History:			Oven dried (100 deg.)
Preparation Method:			Dry sieved
Crumbling / Curling of linear shrinkage:			Nil
Linear shrinkage mould length:			250mm

Table 1.4 Particle size distribution and consistency limits test report (sand from Pit 1).

Client:	Home PTY LTD	Job No.	97638247
Project:	Riverside Development – Pimpama	Date Checked:	19-Jan-2009
Location:	Pit 1: Refer to Map	Checked by:	

Lab Reference No.	B4577	Sample Identification:	Pit 1: 1.0 – 1.5 m

Laboratory Specimen Description: Sand, fine to coarse
Sample size: Diameter 50 mm, height 100 mm

Particle Size Distribution AS1289 3.6.1		Consistency Limits and Moisture Content			
Sieve Size, mm	**% Passing**	**Test**		**Method**	
19.0		**Liquid Limit**	%	AS1289 3.1.2	ND
13.2		**Plastic Limit**	%	AS1289 3.2.1	ND
9.5		**Plasticity Index**	%	AS1289 3.3.1	ND
6.7		**Linear Shrinkage**	%	AS1289 3.4.1	ND
4.75	100	**Moisture Content**	%	AS1289 2.1.1	17.2
2.36	98	**Wet Density**	t/m³	AS1289 F4.1	1.72
1.18	96	**Soil Particle Density**	t/m³	AS1289 F4.1	2.52
0.600	90	**Maximum void ratio**			0.932
0.425	74	**Minimum void ratio**			0.527
0.300	38				
0.150	16	Sample History:			Oven dried (100 deg.)
0.075	3	Preparation Method:			Dry sieved
		Crumbling / Curling of linear shrinkage:			Nil
		Linear shrinkage mould length:			250mm

Question: *The project description contains lots of information, what shall we start with?*
Answer: Geotechnical projects are typically divided into the following steps: a) collection, interpretation and analysis of field data; b) laboratory testing of soil samples obtained from the site and establishment of soil properties; c) identification of potential issues that can occur during or after construction and d) recommendation of most appropriate methods to deal with potential geotechnical issues.

As soil mechanics deals with underground soil conditions, it is necessary to have a good understanding of what soil or rock types are located at different depths. To do this, we will analyze the data from field investigation (borehole logs and DCP tests) and draw a soil cross-section (also known as soil profile) of the site. By interpreting field data, we will identify the main geological units and detect the ground water level. Chapter 2 will deal with basic geology and soil formation and discuss important aspects of field investigation.

Question: *Soil mechanics seems to be a hands-on discipline; can we learn how to conduct field investigation from this book?*
Answer: Unfortunately, no book can give you practical skills necessary to conduct field or laboratory investigations; it is something that you learn and develop by doing yourself. This book will show you why you need to study soil mechanics and help you learn and practice how to interpret and analyze data from field investigations and laboratory tests so that you can easily identify and deal with potential natural hazard or geotechnical issues. To find out more about field and laboratory tests, we recommend viewing videos of relevant tests, which are readily available on the Internet.

Chapter 2

Soil formation and exploration

Project relevance: In this chapter, we will become familiar with data from field investigation and learn how to interpret borehole logs and draw a cross-section. The knowledge of geological settings of the study area gives us a better understanding of the dominant soil type, its spatial variations and ground water conditions. It also allows us to better assess geological and geotechnical problems that may occur in different soil types during or after construction.

2.1 Rock weathering and soil formation

Soils are a product of rock weathering. Although rocks are hard materials, they tend to lose their strength and deteriorate due to weathering over a geological period of time (i.e., hundreds/millions of years). There are two types of weathering: mechanical (or physical) and chemical.

Mechanical weathering (for example, heating and cooling of rocks) is the breakdown of rocks into smaller particles. *Chemical* weathering is the breakdown of rock by chemical reaction that typically involves the effect of water and environment. Heavily weathered rocks eventually transform into soil, a process that may undermine the stability of natural slopes (Kim *et al.*, 2015; Cogan *et al.*, 2018). As the strength of soil is significantly lower than that of rock, heavily weathered material, as shown in Figure 2.1, may not be able to hold the steep slope anymore during rainfalls.

Figure 2.1 Landslide in weathered rock on Gold Coast (Australia). The rock is heavily weathered and partially disintegrated into soil.

Unlike rocks, soil consists of much smaller-sized material (gravel, sand, silt and/or clay), forming two main groups: coarse-grained soils (sand and gravel) and fine-grained soils (predominantly silt and clay).

2.2 Residual and transported soils

After soil is formed, it can remain in the place of its origin (such soil is known as "residual" soil) for many years or it can be transported to new places ("transported" soil). Transported soils are more common (and better studied) and they are commonly categorized based on the agent of transportation.

Question: *Why do we need to know soil origin? Every soil consists of the same components such as gravel, sand, silt and clay, then what is the origin for?*
Answer: Although soils generally consist of the same components, they often have different soil structures (i.e., the way soil particles are arranged against each other), which are generally related to soil origin. As different soil structures result in different soil properties, the knowledge of soil origin can provide us with important information about what geotechnical problems we may have to deal with.

2.2.1 Transported soils

Glacial soils: formed by transportation and deposition of glaciers. This type of soil is typically of high strength and it generally provides a high bearing capacity for engineering structures.

Alluvial soils: transported by running water and deposited along water streams (rivers). They are typically soft and loose material saturated with water. Many geotechnical problems including liquefaction and large settlements of engineering structures occur in alluvial soils.

Marine soils: formed by deposition in the sea or ocean. This type of soil tends to be rather soft and saturated (Fig. 2.2). It is mostly fine-grained material such as silt and clay. Similar to alluvial material, marine soils can undergo significant settlements under stresses.

Figure 2.2 Very soft and saturated mud from the Port of Brisbane (Australia). It undergoes large settlements under stresses.

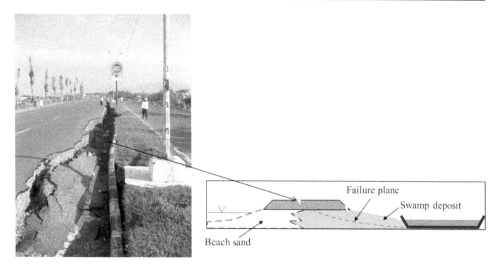

Figure 2.3 Lateral spread and cracks in the pavement (photo) observed at sea dike in Sumatra (Indonesia) caused by the Sumatra earthquake of September 30, 2009. Part of the dike was built on soft swamp deposits which underwent large deformations during the earthquake, resulting in pavement damage (Gratchev *et al.*, 2011).

Figure 2.4 Village was built on an ancient landslide (a) which resulted in collapse of several buildings (b) during the 2008 Sichuan earthquake in China.

Lacustrine soils: formed by deposition in lakes and have similar properties as marine soils. This type is not widely spread as it is limited to the area of lakes and swamps (Fig. 2.3).

Aeolian soils: transported and deposited by wind. It is typically of yellow color and made of fine (mostly silt) particles. Although this soil can be very hard when dry, it may quickly *collapse* when it becomes moist.

Colluvial soils: formed by movement of soil from its original place due to gravity, for example landslide mass (Fig. 2.4). It is a loose deposit, not consolidated and of low strength.

2.2.2 Project analysis: soil origin

Examination of borehole logs (Figs. 1.2–1.4) reveals two major types of soil in the project area: alluvial and residual soil. The presence of very soft and wet alluvium (this soil layer is about 8 m thick in BH1) raises geotechnical concerns as this soil type can potentially cause lots of trouble due to its "unfavorable" properties. The alluvium in BH1, BH6 and BH9 is described as very soft ("VS" in the "consistency" column) and wet ("W" in the "moisture content" column). In contrast, the residual soil is predominantly firm (F) and only moist (M), which makes it better foundation for construction. We will discuss the properties of these soils in detail in the following chapters.

Question: *What shall we do about the topsoil, is it important?*
Answer: Geotechnical engineers are not usually interested in the properties of the top meter of soil (known as topsoil), in which plants grow. It is often not suitable for use as an engineering material because it has high organic content and it is too variable in character.

2.3 Soil mineralogy

When rocks undergo the weathering process, their mineral composition changes; i.e., rock minerals turn into secondary or clay minerals. There are a few common clay minerals such as kaolinite, illite and smectite (or montmorillonite) that play an important role in soil behavior because they determine the properties such as plasticity, permeability, compressibility and strength (Gratchev and Towhata, 2009, 2016). Soils with smectite or montmorillonite frequently cause problems during construction because such soils can absorb and hold lots of water, which would make them loose and weak.

Question: *Is it essential for each project to determine the clay mineralogy?*
Answer: It is advisable but not a must. It is not very common for small to medium-scaled projects in which more attention is paid to soil characteristics such as grain size distribution, plasticity and strength, which are used for design purposes. In addition, special equipment is required to perform X-ray tests on soil, which makes it relatively expensive.

2.4 Soil exploration

Soils are very complex materials and vary widely. There is no certainty that soil will have the same properties within a few centimeters of its current location. Soils exploration is conducted only on a fraction of a proposed site because it would be very expensive to conduct an extensive investigation of the whole construction site. For this reason, engineers are required to make estimates and judgments based on the information from a limited set of observations and laboratory test data. Field investigation and laboratory tests must be performed in accordance with the relevant standards (Table 2.1).

2.4.1 Desk study

Before any field and laboratory work, engineers and geologists are required to conduct a "desk study" during which they collect as much relevant information as possible, which may include maps (geological, seismic and hydrological), existing literature or geotechnical reports related to the area of investigation. Geological maps provide important information on site geology and topography (Fig. 2.5) and they are essential in planning of effective site investigations.

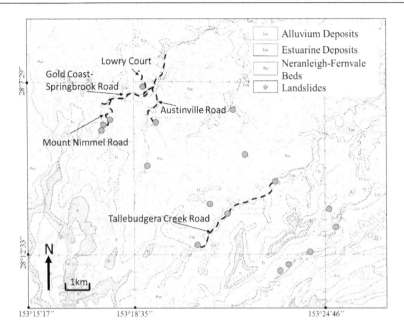

Figure 2.5 Geological map of the Springbrook-Tallebudgera part of Gold Coast (Australia). This map indicates the location of several landslides triggered by the tropical cyclone Debbie in March 2017, making it clear that most of landslides occurred in the deposits of Neranleigh-Fernvale Beds (More details are given in Cogan et al., 2018).

Question: *The geological map in Figure 2.5 provides information about the surface geology. How do we know what to expect at different depths, for example 2–5 m below the ground surface?*
Answer: Each geological map typically includes a geological cross-section that provides some information about the geological units with increasing depths. Without such a cross-section, the geological map becomes less useful.

There are different ways of investigating the site conditions, which are documented in the relevant standard codes. Depending on the scope of investigation and available budget, the field work can include boreholes, field testing such as standard penetration tests (SPT), dynamic cone penetration (DCP) tests and/or geophysics surveys.

2.4.2 Boreholes

Boreholes are considered the most effective but also most expensive method of site investigation. As they can be drilled to different depths, boreholes provide essential data about soil type and soil properties. In addition, soil and/or rock core (Fig. 2.6) is collected from boreholes for examination and laboratory testing.

Borehole data help engineers and practitioners to better understand the geological settings of the area, detect the boundaries between different soil layers and obtain soil properties. It also contains important information about soil *in-situ* properties such as consistency and moisture. Using the borehole data, engineers draw a soil cross-section that shows type of soil and its location with depths.

Figure 2.6 Borehole samples (a) collected from a depth of 70 m in the Aratazawa landslide site (Japan). They were transported to a geotechnical laboratory and tested to determine the soil strength characteristics. The Aratazawa landslide was one of the largest landslides in the past century with a landslide body length of more than 1 km (Gratchev and Towhata, 2010). The height of the newly created cliff was almost 150 m (b).

Question: *How close to each other and how deep should boreholes be?*
Answer: The distance between boreholes and their depth depend on the type of project and available budget. This information is provided in the relevant standards on geotechnical investigation, which suggest that the number of boreholes should increase: a) as soil or rock variability increases; b) as the loads from engineering structures increase; c) for more important structures.

Unfortunately, boreholes provide data only for one point in space (i.e., where they are drilled), which, in many cases, may not be sufficient to characterize a much larger area of investigation.

Question: *If boreholes are spread apart, how do we know what soil type is in between them?*
Answer: We don't know, that's why we often have to make assumptions based on our knowledge of site geology and common sense. This practice can be risky as there have been many examples where wrong soil boundaries were assumed or thin layers of very soft clay were not detected during investigation. This may lead to severe consequences including structure damage or even collapse. To avoid this, we should also use other methods of site investigations such as geophysics or field tests including standard penetration tests (SPT) and cone penetration tests (CPT). Although these tests do not provide information about the soil type, they give important data on soil properties such as strength.

2.4.3 Laboratory tests

The borehole samples are transported from the field site to the laboratory for more detailed examination. It typically includes sieve and Atterberg limits tests, which are performed by all geotechnical companies. Additional laboratory investigation such as standard Proctor compaction tests and constant head (or falling head) tests will provide important information about soil compaction characteristics and its hydraulic conductivity, respectively. Compressibility and strength of soil can be examined by means of oedometer and shear box tests (or triaxial tests), respectively. Table 2.1 summarizes geotechnical standards used in different countries for soil testing.

Table 2.1 List of tests and applicable standards.

Test	Applicable standards		
	ASTM (USA)	AS Australian standard	BS British standard
Site investigation	D5922, D5923, D5924	1726	5930
Moisture content of soil	D2216	1289.2.1.1	1377:Part 2:1990
Analysis of grain size distribution	D422, D1140	1289.3.6.1	1377:Part 2:1990
Specific gravity of soil solids	D854	1289.3.5.1	1377:Part 2:1990
Liquid limit and plastic limit of soil	D4318	1289.3.1.1, 1289.3.2.1	1377:Part 2:1990
Laboratory classification of soil	D2488	2870:2011 Site classification	5930
Laboratory soil compaction	D698, D1557	1289.5.1.1	1377:Part 4:1990
Hydraulic conductivity of granular soil	D2434	1289.6.7.1, 1289.6.7.2	1377:Part 5:1990
One-dimensional consolidation test of cohesive soil	D2435	1289.6.6.1	1377:Part 5:1990
Direct shear strength test of granular soil	D3080	1289.6.2.2	1377:Part 7:1990
Unconfined compressive strength test	D2166	1289.6.4.1	1377:Part 7:1990

2.5 Project analysis: field and laboratory data

2.5.1 Borehole logs and cross-section

Soil cross-sections are important in geotechnical investigations because they indicate the geological conditions that exist at the construction site, including the type of soil, boundaries between different units, and ground water conditions. A cross-section through A-A' line from Figure 1.1 is given in Figure 2.7.

Question: Why are the layers connected with dotted lines in Figure 2.7?
Answer: The cross-section provides subsurface information, which is, however, limited to the point where the borehole was drilled. Although it is common to assume that soil layers are spread in a horizontal direction, it is still an assumption. We use dotted lines because we do not know exactly what soil conditions exist between the boreholes. It is also acceptable not to connect the soil layers from different boreholes.

It is evident from this cross-section that the subsurface conditions at the project site consist of topsoil (about 0.5 m) overlying a stratum of soft and wet alluvial silty clay (alluvium). The firm to stiff residual clay is found at depths of 5 to 8 m.

Question: I do not have much experience in drawing cross-sections, are there any guidelines I can follow?
Answer: There may be different approaches in drawing geotechnical cross-sections. However, we are generally required: a) to provide the legend (when necessary); b) to spell out all symbols, color and patterns; c) to use dotted lines to connect soil layers

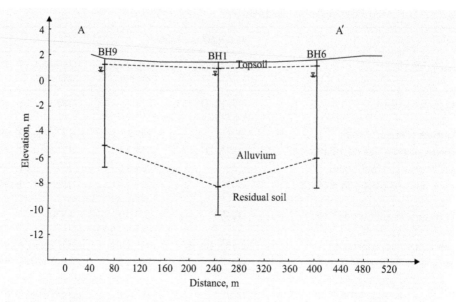

Figure 2.7 A cross-section through the A-A' line in Figure 1.1.

(or do not connect the soil layers at all); d) not to connect the end of boreholes because the end of boreholes doesn't mean the end of the soil layer; e) to provide the scale and units for each axis.

To study the properties of alluvium and residual soil, several undisturbed soil samples were collected at different depths (the depth is indicated as U50 in the "Sample type" column of each borehole log) and examined in the laboratory. The data from index properties tests are summarized in "Particle size distribution and consistency limits test" reports (Tables 1.2–1.4).

Question: *In the "additional observations" column of each borehole log (Figs. 1.1–1.3), there is information about "pp", against each U50 symbol, for example, pp 15–20 kPa. What does it mean and how can we use it?*
Answer: The abbreviation "pp" stands for pocket penetrometer, which is a device that can quickly provide an estimation of unconfined compressive strength (q_u, in kPa). According to Look (2014), for general material $q_u \approx 0.8$ pp while for fissured clays it can be as $q_u \approx 0.6$ pp. Although this test is fast and simple, it is considered not very reliable, and care needs to be taken when using such data. In the case of borehole data analysis, pp values can be used to determine the boundaries between different units such as the relatively weak alluvium (average pp values of 15–20 kPa) and the relatively hard residual soil (pp values are more than 100 kPa).

2.5.2 Field work: dynamic cone penetration test

A portable dynamic cone penetration test was performed near Borehole BH9 to obtain estimates of the strength of soil and to clarify the boundaries between the geological units. The penetrometer (Fig. 2.8) had a conical tip with a 4.9 cm² area and 60° apex angle. It was driven in the ground by repeated blows of a 5 kg hammer released from a height of 50 cm.

The number of blows required to achieve 10 cm of penetration was taken as the cone penetration resistance (N_d) and summarized in Table 1.1.

We will plot the data from Table 1.1 in Figure 2.9 to study the variation of soil resistance to penetration with depths. As can be seen in this figure, the soil resistance is relatively low

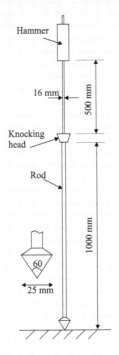

Figure 2.8 Schematic illustration of dynamic cone penetrometer (DCP).

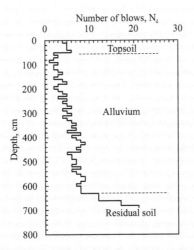

Figure 2.9 Results and interpretation of dynamic cone penetration test conducted near BH9.

at depths of 0.5 to 6.5 m (N_d varies from 2 to 8) which corresponds to the depths of very soft alluvial clay. The resistance increases below the depth of 6.5 m, indicating the presence of a harder layer of soil (which appears to be the residual soil, according to the BH9 log).

Question: *How deep can we go with this test?*

Answer: In general, we can penetrate about 2–3 m for most soils and sometimes go even deeper (5–6 m) for very soft material. However, it may become technically difficult to retract the rod from the ground when the depth of penetration is too large.

2.6 Review quiz

1. The two weathering processes are

   a) water and air   b) physical and chemical
   c) streams and rivers   d) human and nature

2. What statement is NOT correct?

   a) The exposed rocks at the surface of the earth are subject to weathering
   b) Weathering causes rocks to become more porous
   c) Volcanic rocks are more resistant to weathering than other types of rock
   d) Soils are the product of rock weathering

3. What soil is typically associated with "collapsibility"?

   a) lacustrine   b) aeolian   c) alluvium   d) marine

4. What type of soil is formed by deposition in lakes?

   a) lacustrine   b) alluvial   c) marine   d) aeolian

5. Which of the following is not a clay mineral?

   a) montmorillonite   b) smectite   c) chlorite   d) feldspar

6. Liquefaction is typically associated with

   a) gravel   b) sand   c) silt   d) clay

7. The first phase of geotechnical investigation is to

   a) collect all available information
   b) make a site visit
   c) prepare borehole locations
   d) all of above (a-c) need be done simultaneously

8. Alluvial soils are soils transported by

   a) ice   b) river   c) wind   d) humans

9. Expansive soils swell because they contain an abundance of

    a) kaolinite   b) illite   c) montmorillonite   d) quartz

10. In general, what type of soil would have more favorable geotechnical characteristics and cause fewer geotechnical issues?

    a) glacial   b) alluvium   c) marine   d) colluvium

Answers: 1) b   2) c   3) b   4) a   5) d   6) b   7) a   8) b   9) c   10) a

Chapter 3

Soil constituents

Project relevance: Once we have established the main geological units (i.e., soil layers) using the borehole data, we will proceed with determining the basic soil properties, which will help us better understand what problems can occur during construction. The following chapter will explain how to characterize soil using a three phase diagram.

3.1 Three phases in soil

Soil consists of three major phases such as *solid particles, water and air*. Solid or soil particles create voids that can be filled with water and/or air as shown in Figure 3.1a. To determine the basic properties of soil, we should know the amount of each component. For this reason, a soil sample is divided into three phases as schematically shown in Figure 3.1b in regards to soil volume and mass. The next section will introduce important definitions related to volume (volumetric ratios) and mass (mass ratios).

Question: *Do "mass" and "weight" mean the same thing in soil mechanics?*
Answer: In our daily life, mass and weight often mean the same thing; however, in soil mechanics, they are different. Unlike the term mass (which is measured in kilograms), weight is force measured in Newton (N).

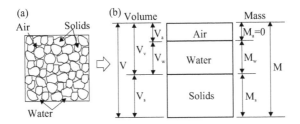

Figure 3.1 Three phases in soil. Volume components: V is the volume of soil sample, V_s is the volume of solids, V_v is the volume of voids, V_w is the volume of water and V_a is the volume of air. Mass components: M is the mass of soil sample, M_s is the mass of solids, M_w is the mass of water and M_a is the mass of air ($M_a = 0$).

3.2 Volumetric ratios

3.2.1 Void ratio

Soil contains lots of void space, which makes it rather porous and soft, compared to rocks. *Void ratio* (e) indicates the amount of voids in soil and it is defined as the ratio of the volume of void space (V_v) to the volume of solids (V_s) (Equation 3.1). The higher the void ratio, the looser the soil structure is. Soil with higher void ratios will likely cause more problems during construction as it may undergo larger settlements.

$$e = \frac{V_v}{V_s} \tag{3.1}$$

Void ratio seems to be the most common "volumetric" parameter used in soil mechanics. On average, the void ratio of sand varies from 0.6 to 1.0, while for clay it is typically higher, ranging from 0.8 to 2.2. For organic soils, the void ratio can be as high as five. For engineering purposes, it is always desirable to deal with soil that has relatively low values of void ratio.

Question: *Is void ratio presented as percentage (for example, e = 60%) or ratio (e = 0.6)?*
Answer: The void ratio is always given as a dimensionless number, i.e., e = 0.6 is correct while e = 60.0% is never used.

3.2.2 Relative density

When we collect a soil sample during field investigation, we measure its *in-situ* void ratio (e_0). If this sample is compacted to its densest state, its void space will decrease and the void ratio will drop to its minimum value (e_{min}). It is also possible to achieve the loosest state for this soil in the laboratory so that its void ratio will reach its maximum value of e_{max} (Fig. 3.2).

The relative density (D_r) is used to compare the *in-situ* state of soil with its loosest and densest states (Equation 3.2).

$$D_r(\%) = \frac{(e_{max} - e_0)}{(e_{max} - e_{min})} 100\% \tag{3.2}$$

When soil has a low value of the relative density (D_r) it can be classified as "very loose" or "loose" (Table 3.1). Such soil would typically have loose structures and can be more susceptible to failure under loading. Relative density is frequently used for coarse-grained material to estimate its liquefaction potential.

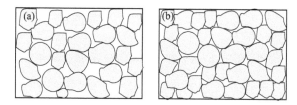

Figure 3.2 Soil structure with different void space: (a) loosest state (e_{max}), and (b) densest state (e_{min})

Table 3.1 Qualitative description of soil deposits.

Relative density (%)	Soil state
0–15	Very loose
15–35	Loose
35–65	Medium
65–85	Dense
85–100	Very dense

Question: *The relative density seems to be mostly used for sand. What about clay?*
Answer: For clay, plasticity has a much stronger effect on soil properties, and for this reason, it is typically used as an indicator of soil resistance to liquefaction (Gratchev et al., 2006).

3.2.3 Porosity

Porosity (n) refers to the voids of soil as well; however, it is more common for a rock-soil type. It is the ratio of the volume of voids (V_v) to the total volume of soil (V) (Equation 3.3) and it can be reported in percentage as well (for example, n = 34%).

$$n = \frac{V_v}{V} = \frac{e}{1+e} \qquad (3.3)$$

3.2.4 Degree of saturation

The void space in soil can be empty (i.e., filled with air) or filled with water. The degree of saturation (S) indicates how much water is present in the voids. It is defined as the ratio of the volume of water (V_w) to the volume of void space (V_v) (Equation 3.4). Soil whose void space is filled with water is called *saturated* and its degree of saturation equals 1 (or 100%). For dry soil, the void space is filled with air and S = 0.

$$S = \frac{V_w}{V_v} \qquad (3.4)$$

In many textbook examples, it is commonly assumed that the soil below the ground water table is saturated while the soil above the ground water table is partially saturated or dry. In real life, soil below the ground water table may not be fully saturated as there is always some air trapped in the soil.

Question: *If the soil is partially saturated, let's say S = 0.75, what shall we call it?*
Answer: It is referred to as "unsaturated soil". Behavior of unsaturated soil is more complex and less studied than the behavior of saturated soils. It has been shown that suction can be developed in unsaturated soils due to capillary forces and it can provide additional strength to soil. However, in many design codes, the effect of suction is generally neglected and design is performed for saturated soil, which is generally considered to be the worst-case scenario. More information about unsaturated soils can be found in Fredlund and Rahardjo (1993).

3.3 Mass ratios

3.3.1 Density

Density (ρ) is the ratio of the mass of soil sample (M) to its volume (V). If soil contains water, its total mass will include the mass of solid particles (M_s) and the mass of water (M_w) (Equation 3.5). Note that different terms such as *bulk, total, wet* or *moist* density are used in the literature to refer to soil density.

$$\rho = \frac{M}{V} = \frac{M_s + M_w}{V}$$

(3.5)

Remember that the *density of water* (ρ_w) is commonly taken as 1 g cm^{-3} = 1 t m^{-3} = 1000 kg m^{-3}.

Dry density is the ratio of the mass of solids (M_s) to the soil volume (V) (Equation 3.6). The dry density is commonly used to analyze data from Proctor compaction tests (see Chapter 5).

$$\rho_d = \frac{M_s}{V}$$

(3.6)

Question: *Do the dry density and density of solids mean the same thing?*
Answer: No, they are two different properties. The density of solids (also known as the density of solid particles) is the mass of solids (i.e., mass of dry soil) divided by the volume of solids (V_s) while the dry density is the ratio between the mass of solids and the total volume (V) of soil. The difference is in the volumes (V_s against V) used. As the volume of solids is smaller than the total volume of soil, the density of solids is always greater than the dry density of soil. For your reference, the density of solids commonly ranges between 2.6 and 2.8 g cm^{-3}, while the dry density of soil varies from 1.1 to 1.8 g cm^{-3}, depending on soil conditions.

3.3.2 Unit weight

The unit weight (γ) of soil is the ratio between the soil weight (W) and its volume (V). It is related to soil density (ρ) as shown in Equation 3.7. Similar to soil density, the unit weight can be referred to as "bulk unit weight", "moist unit weight" or "dry unit weight", depending on the presence of water in soil. The common range of unit weight for soils is 14–22 kN m^{-3}, depending on soil structures and water content.

$$\gamma = \frac{W}{V} = \rho \cdot g$$

(3.7)

where g is the gravitational acceleration ($\approx$9.81 m s^{-2}).

Question: *Why do we need to know the unit weight, isn't just the soil density enough?*
Answer: In soil mechanics, the unit weight of soil is used to calculate stresses acting in soil mass (see Chapter 6) and for this reason, the unit weight is more commonly used than soil density.

3.3.3 Water content

The water (or moisture) content (w) indicates the amount of water in soil. It is the ratio of the mass of water (M_w) to the mass of solids (M_s) (Equation 3.8). For saturated fine-grained soils, the water content can exceed 100% while for some very plastic soils that contain clay minerals such as smectite or montmorillonite, it can reach more than 700%. High water content is another indicator that soil may be loose and weak, and thus special attention needs to be given to this soil.

$$w = \frac{M_w}{M_s} \tag{3.8}$$

Question: *Can all soils have water content more than 100% or is it only possible for plastic soils?*
Answer: Very high water content is common for high plasticity soils that contain smectite or montmorillonite minerals. The clay particles of such clays are extremely small, forming very loose structures with lots of voids (i.e., very high void ratios). In addition, due to the specific structure of these clay minerals, a negative charge exists on their sides that can attract and hold water. Gravel and sand have relatively larger solid particles and they do not have this ability to attract and hold water. For this reason, coarse-grained soils generally have water content around 30–40% when saturated. Please note that organic soils can also have very high water content.

3.3.4 Specific gravity

Specific gravity (G_s) of solids is defined as the ratio between the density of solid particles and the density of water (Equation 3.9). It indicates how heavy the soil particles are compared to water. The specific gravity is a dimensionless number, which typically varies from 2.6 to 2.8. As many soils contain a significant amount of quartz, their specific gravity would be close to the specific gravity of quartz (about 2.7).

$$G_s = \frac{\rho_s}{\rho_w} \tag{3.9}$$

where ρ_s is the density of solid particles and ρ_w is the density of water.

Question: *Why is the range of G_s values very narrow?*
Answer: The range is narrow because many types of soil contain significant amounts of quartz. For this reason, their specific gravity would be close to the specific gravity of quartz, which is about 2.65–2.7.

3.4 More about soil constituents

There are a few more commonly used equations which are related to dry density (ρ_d) (Equation 3.10), unit weight (γ) (Equation 3.11), saturated unit weight (γ_{sat}) (Equation 3.12) and degree of saturation (S) (Equation 3.13).

$$\rho_d = \frac{\rho}{1+w} = \frac{G_s \rho_w}{1+e} = \frac{G_s \rho_w (1 - A_v)}{1 + w G_s} \tag{3.10}$$

where A_v is the air content defined as the ratio between the volume of air (V_a) and total volume of soil (V).

$$\gamma = \frac{G_s\left(1+w\right)}{1+e}\gamma_w \tag{3.11}$$

$$\gamma_{sat} = \frac{G_s+e}{1+e}\gamma_w \tag{3.12}$$

$$S = \frac{w \cdot G_s}{e} \tag{3.13}$$

Table 3.2 summarizes the most common equations and typical ranges of soil basic properties.

Table 3.2 Typical range of soil properties.

Parameter	Equation	Typical range
Density	$\rho = \dfrac{M}{V}$	1.4–2.2 g cm^{-3}
Dry density	$\rho_d = \dfrac{M_s}{V}$	1.2–2.0 g cm^{-3}
Unit weight	$\gamma = \dfrac{W}{V}$	14–22 kN m^{-3}
Dry unit weight	$\gamma_d = \dfrac{W_s}{V}$	12–20 kN m^{-3}
Water content	$w = \dfrac{M_w}{M_s} \cdot 100\%$	10–50%
Void ratio	$e = \dfrac{V_v}{V_s}$	0.4–1.5
Porosity	$n = \dfrac{V_v}{V} \cdot 100\%$	25–60%
Degree of saturation	$S = \dfrac{V_w}{V_v} \cdot 100\%$	10–100%
Specific gravity	$G_s = \dfrac{\rho_s}{\rho_w} = \dfrac{W_s}{\gamma_w \cdot V_s}$	2.65–2.80

3.5 Project analysis: soil constituents

Let's apply our knowledge of soil constituents to determine basic properties of soil from the project site. We will refer to the data from Tables 1.2–1.4 and select the sandy soil from Pit 1 (Table 1.4) as an example.

Solution for the sand sample

The volume of sand sample is calculated as

$$V = \pi r^2 h = 196.3\,\text{cm}^3$$

Specific gravity equals

$$G_s = \frac{\rho_s}{\rho_w} = \frac{2.52}{1} = 2.52$$

Using Equation 3.10, we find the dry density as

$$\rho_d = \frac{\rho}{1+w} = \frac{1.72}{1+0.172} = 1.47 \text{g cm}^{-3}$$

Using the definition of density, void ratio and degree of saturation, we will find:

Mass of soil, $M = \rho \cdot V = 1.72 \cdot 196.3 = 337.6 \text{g}$

Mass of dry soil, $M_d = \rho_d \cdot V = 1.47 \cdot 196.3 = 288.6 \text{g}$

Mass of water, $M_w = M - M_d = 49.0 \text{g}$

Volume of water, $V_w = \frac{M_w}{\rho_w} = \frac{0.049}{1000} = 0.000049 \text{m}^3$

Volume of solids, $V_s = \frac{M_s}{\rho_s} = \frac{288.6}{2.52} = 114.5 \text{cm}^3 = 0.000114 \text{m}^3$

Volume of air,
$V_a = V - V_s - V_w = 0.000196 - 0.000114 - 0.000049 = 0.000032 \text{m}^3$

Void ratio, $e = \frac{V_w + V_a}{V_s} = 0.718$

Degree of saturation, $S = \frac{V_w}{V_w + V_a} 100\% = 60.4 \%$

Weight of solids, $W_s = \frac{M_s}{101.97} = 2.82 \text{N}$

Using Equation 3.2, the relative density will be

$$D_r = \frac{(e_{max} - e_0)}{(e_{max} - e_{min})} 100\% = \frac{(0.932 - 0.718)}{(0.932 - 0.527)} 100\% \approx 53\%$$

The results obtained for the alluvial and residual soil are summarized in Table 3.3. Note that D_r is mostly used for the non-plastic sand.

Question: *There are too many variables in Table 3.3, which are the most important?*
Answer: We are mostly interested in soil density, void ratio and degree of saturation because these soil properties can give us a good idea about the engineering behavior of soil. For example, the void ratio of the silty clay (alluvium) is very high (e = 2.07), and it is significantly higher than the void ratio of the sand and residual soil. It implies that the alluvial material is very loose. In addition, its degree of saturation is very close to 100% (S = 95.7%), which means that the pore space is mostly filled with water. This

Table 3.3 Geotechnical characteristics of soils from the project site.

	Silty Clay (Alluvium)	Sand (Pit 1)	Sandy Silty Clay (Residual soil)
Sample size: *diameter 50 mm, height 100 mm*			
Depth (m)	2.6–3.0 m	8.2–8.6 m	10.1–10.5 m
Maximum void ratio	N/A	0.932	N/A
Minimum void ratio	N/A	0.527	N/A
Wet density (g cm⁻³)	1.47	1.72	1.95
Water content (%)	78.3	17.2	19.1
Dry density (g cm⁻³)	0.82	1.47	1.64
Specific gravity	2.53	2.52	2.63
Volume of water (m³)	0.000127	0.0000495	0.000061
Weight of solids (N)	1.59	2.82	3.15
Void ratio	2.07	0.718	0.607
Degree of saturation (%)	95.7	60.4	82.8
Relative density (%)	N/A	53.1	N/A

clearly presents unfavorable conditions for construction, as this soil will likely undergo significant deformations under loads. The void ratio of the other two soil types (sand and residual soil) is relatively low, while the dry density is high, making them more suitable for construction purposes. In conclusion, the unfavorable geotechnical properties of the silty clay (alluvium) may cause problems as the project continues. Prior to construction, the properties of this soil must be improved so that its void ratio and degree of saturation would decrease to an acceptable level.

3.6 Problems for practice

Problem 3.1 A soil sample was collected for laboratory examination. It has a wet mass of 5.2 kg, bulk density of 1.65 g cm⁻³, dry density of 1223 kg m⁻³ and degree of saturation of 82%. Determine the density of solids.

Solution

There are different ways to solve this problem; we will use the definitions of soil constituents. We will first find the mass of solids (i.e., the mass of dry soil), then the volume of solids and finally its density.

$$\text{Volume of soil sample, } V = \frac{M}{\rho} = \frac{5.2}{1650} = 0.00315\,\text{m}^3$$

$$\text{Mass of dry sample, } M_d = \rho_d \cdot V = 1223 \cdot 0.00315 = 3.85\,\text{kg}$$

$$\text{Mass of water, } M_w = M_{soil} - M_d = 5.2 - 3.85 = 1.35\,\text{kg}$$

$$\text{Volume of water, } V_w = \frac{M_w}{\rho_w} = \frac{1.35}{1000} \approx 0.00135\,\text{m}^3$$

$$\text{From } S = \frac{V_w}{V_v} = 0.82$$

We will obtain the volume of voids as

$$V_v = \frac{V_w}{S} = \frac{0.00135}{0.82} = 0.00164\,\text{m}^3$$

Then, the volume of solids equals

$$V_s = V - V_v = 0.00315 - 0.00164 \approx 0.0015\,\text{m}^3$$

Therefore, the density of solids is

$$\rho_s = \frac{M_d}{V_s} = \frac{3.85}{0.0015} = 2{,}566\,\text{kg m}^{-3}$$

Problem 3.2 Site investigation was performed to study soil conditions at a construction site in a new development area. A cylindrical soil sample (height = 100 mm, diameter = 50 mm) was collected at a depth of 1.5 m below the ground. The following soil characteristics were obtained: soil density was 1.52 t m^{-3}, moisture content was 68.2% and density of solid particles was 2.53 g cm^{-3}. Determine:

a) Weight of solids (in N)
b) Volume of air (in m^3).

Solution

Similar to Problem 3.1, there are different ways to solve it; we will use the definitions of soil constituents.

Volume of soil sample, $V = \pi r^2 h = 0.000196\,\text{m}^3$

Unit weight of soil, $\gamma = 1.52 \cdot 9.81 = 14.9\,\text{kN m}^{-3}$

Weight of soil, $W = \gamma \cdot V = 14.9 \cdot 1000 \cdot 0.000196 = 2.92\,\text{N}$

Weight of solids, $W_s = \dfrac{W}{1+w} = \dfrac{2.92}{1+0.682} \approx 1.74\,\text{N}$

Specific gravity, $G_s = \dfrac{\rho_s}{\rho_w} = \dfrac{2.53}{1} = 2.53$

Volume of solids, $V_s = \dfrac{W_s}{\gamma_s} = \dfrac{W_s}{G_s \cdot \gamma_w} = \dfrac{1.74}{2.53 \cdot 9.81 \cdot 1000} \approx 7 \cdot 10^{-5}\,\text{m}^3$

Weight of water, $W_w = W_s \cdot w = 1.74 \cdot 0.682 = 1.19\,\text{N}$

Volume of water, $V_w = \dfrac{W_w}{\gamma_w} = \dfrac{1.19}{9.81 \cdot 1000} \approx 0.00012\,\text{m}^3$

Volume of air, $V_a = V - V_s - V_w \approx 5.2 \cdot 10^{-6}\,\text{m}^3$

Problem 3.3 Soil excavated from a borrow pit is being used to construct an embankment (Fig. 3.3). The soil sample from the borrow pit has a specific gravity of 2.7 and unit weight of 17.8 kN m^{-3}. The weight of the sample was 3.5 N. The sample was then placed in an oven for 24 h at 105°C and its weight reduced to a constant value of 2.9 N.

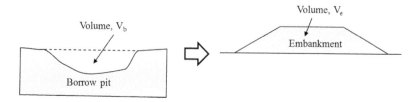

Figure 3.3 Borrow pit and embankment for Problem 3.3.

The soil at the embankment is required to be compacted to a void ratio of 0.71. If the finished volume of the embankment (V_e) is 80,000 m³, what would be the volume of the soil (V_b) excavated at the borrow area?

Solution

Please note that there are different ways to solve this problem. This solution will deal with the volume of soil in the embankment (V), in the borrow pit (V_p) and soil void ratios (e_e and e_p, respectively). From the three phase diagram (Fig. 3.1), we can derive that the total volume of soil can be written as V = 1 + e.

Question: *How is the total volume (V) related to the void ratio (e)?*
Answer: For many problems related to soil constituents, it can be assumed that the volume of solids (V_s) is equal to 1 m³ as it makes the solution work-out much easier. Then, from the definition of void ratio (Equation 3.1), the volume of voids (V_v) will be equal to e and thus the total volume of soil will be V = 1 + e.

It can be stated that

$$\frac{V_p}{V_e} = \frac{1+e_p}{1+e_e}$$

To find e_p, the following calculations involving soil water content and unit weight should be done:

Weight of water, $W_w = W - W_s = 3.5 - 2.9 = 0.6\,\text{N}$

Water content, $w = \dfrac{W_w}{W_s} = \dfrac{0.6}{2.9} = 0.21$

Dry unit weight, $\gamma_d = \dfrac{\gamma}{1+w} = \dfrac{17.8}{1+0.21} = 14.7\,\text{kN m}^{-3}$

From $\gamma_{dry} = \dfrac{G_s \cdot \gamma_{water}}{1+e}$

We will find that the void ratio of soil from the borrow pit equals

$e = e_p \approx 0.796$

Finally, we will have

$$\frac{V_p}{80,000} = \frac{1+0.796}{1+0.71}$$

Giving the volume of soil from the borrow pit

$$V_p \approx 84,019\,\mathrm{m}^3$$

Problem 3.4 A 1 m thick soil with the initial void ratio of 0.94 was compacted by a roller and its thickness reduced by 0.09 m (Fig. 3.4). The specific gravity of this soil was 2.65. A 178 g soil sample was collected from the compacted soil mass to examine the degree of compaction; it was dried in an oven for 24 h and it had a dry mass of 142.4 g. Determine the degree of saturation after the compaction.

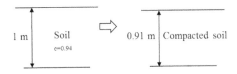

Figure 3.4 Changes in the soil layer thickness in Problem 3.4

Solution

Assume that a width of the soil mass before compaction is 1 m, then its volume is $V = 1\,\mathrm{m}^3$
From the definition of void ratio

$$e = \frac{V_v}{V_s} = \frac{V-V_s}{V_s}$$

We will find the volume of solids as

$$V_s = \frac{V}{1+e} = \frac{1}{1+0.94} = 0.515\,\mathrm{m}^3$$

It is logical to assume that the volume of solids remains the same after the compaction; however, the volume of voids would likely decrease.
    The new volume of the compacted soil mass equals

$$V_{new} = 0.91 \cdot 1 = 0.91\,\mathrm{m}^3$$

Then, the volume of voids will become

$$V_v = V_{new} - V_s = 0.91 - 0.515 = 0.395\,\mathrm{m}^3$$

The new void ratio of compacted soil equals

$$e_{new} = \frac{0.395}{0.515} = 0.765$$

The water content will change to

$$w = \frac{178 - 142.4}{142.4} = 0.25$$

And the degree of saturation will become

$$S = \frac{w \cdot G_s}{e} = \frac{0.25 \cdot 2.65}{0.765} \approx 0.866 \; or \; 86.6 \;\%$$

Problem 3.5 A laboratory specimen of soil has a volume of 2.3 m³. The void ratio of the sample is 0.712 and water content is 16.1%. The specific gravity of the solid particles is 2.7. Determine:

a) Volume of water
b) Mass of solids
c) Dry density
d) Bulk density

Solution

Volume of soil (V) consists of the volume of voids (V_v) and volume of solids (V_s), i.e.,

$$V = V_v + V_s = 2.3\,\text{m}^3 \tag{a}$$

We also know (Equation 3.1) that

$$e = \frac{V_v}{V_s} = 0.712 \tag{b}$$

Substituting V_s from Equation (b) to Equation (a), we get

$$1.712 V_v = 1.64$$

Therefore,

$$V_v = 0.96\,\text{m}^3, \; \text{and} \; V_s = 1.34\,\text{m}^3$$

From $w = \dfrac{S \cdot e}{G_s}$

We will get the degree of saturation (S) as

$$S = \frac{wG_s}{e} = \frac{0.161 \cdot 2.7}{0.712} = 0.61$$

From $S = \dfrac{V_w}{V_v}$

We will find the volume of water

$$V_w = S \cdot V_v = 0.61 \cdot 0.96 \approx 0.59\,\text{m}^3$$

Mass of solids equals

$$M_s = \rho_s \cdot V_s = G_s \cdot \rho_w \cdot V_s = 2.7 \cdot 1000 \cdot 1.34 \approx 3618 \, \text{kg}$$

Dry density (ρ_d) of soil will be

$$\rho_d = \frac{M_s}{V} = \frac{3618}{2.3} = 1573 \, \text{kg/m}^3 = 1.57 \, \text{g/cm}^3$$

Mass of water equals

$$M_w = \rho_w \cdot V_w = 1000 \cdot 0.59 \approx 590 \, \text{kg}$$

From the definition of soil density, we have

$$\rho = \frac{M_s + M_w}{V} = \frac{3618 + 590}{2.3} \approx 1829.6 \, \text{kg/m}^3$$

Problem 3.6 A cylindrical sample of clay, 50 mm (diameter) × 100 mm long, had weight of 3.5 N. It was placed in an oven for 24 h at 105°C. The sample weight reduced to a constant value of 2.9 N. If the specific gravity is 2.7, determine:

a)   Void ratio
b)   Dry unit weight
c)   Degree of saturation

Solution

This problem will be solved using the aforementioned equations/relationships between the soil constituents.

Weight of water, $W_w = 3.5 - 2.9 = 0.6 \, \text{N}$

Bulk unit weight, $\gamma_{bulk} = \dfrac{W}{V} = \dfrac{3.5 \cdot 10^{-3}}{196.4 \cdot 10^{-6}} \approx 17.8 \, \text{kN m}^{-3}$

Water content, $w = \dfrac{W_w}{W_s} = \dfrac{0.6}{2.9} \approx 0.207 \, or \, 20.7 \, \%$

Dry unit weight, $\gamma_d = \dfrac{\gamma}{1+w} = \dfrac{17.8}{1+0.207} \approx 14.7 \, \text{kN m}^{-3}$

From $\gamma_d = \dfrac{G_s}{1+e} \cdot \gamma_w$

We will get that

$$e \approx 0.8$$

Finally, the degree of saturation equals

$$S = \frac{w \cdot G_s}{e} = \frac{0.207 \cdot 2.7}{0.8} \approx 0.7 \, or \, 70 \, \%$$

3.7 Review quiz

1. Which of the following ratios are not related to volume?

   a) degree of saturation   b) water content
   c) void ratio   d) porosity

2. The ratio between the volume of voids and the total volume of soil is called

   a) void ratio   b) porosity
   c) degree of saturation   d) water content

3. What statement is NOT correct?

   a) Water content can be more than 100%
   b) The void ratio of sand is generally greater than the void ratio of clay
   c) The dry unit weight is smaller than the bulk unit weight
   d) The unit weight of soil is measured in kN m^{-3}

4. The density of solids is the ratio between the

   a) mass of bulk (moist) soil and its volume
   b) mass of dry soil and the volume of soil
   c) mass of bulk (moist) soil and the volume of solids
   d) mass of dry soil and the volume of solids

5. Degree of saturation is the ratio between the volume of water and

   a) the total volume of soil   b) the volume of solids
   c) the volume of voids   d) the volume of air

6. The average value of specific gravity (G_s) of clay would be

   a) 1.7   b) 2.7   c) 3.7   d) 4.7

7. What would be the typical value of unit weight (*in kN m^{-3}*) of soil?

   a) 7   b) 17   c) 27   d) 37

8. Water content is the weight of water divided by the

   a) weight of soil   b) weight of solids
   c) mass of soil   d) mass of solids

9. Which of the following values is likely for the mass (*in kg*) of 1 m^3 soil?

   a) 17   b) 170   c) 1700   d) 17000

10. The void ratio of organic soils can be as high as 5.

   a) True   b) False

Answers: 1) b   2) b   3) b   4) d   5) c   6) b   7) b   8) b   9) c   10) a

Chapter 4

Soil classification

Project relevance: There are different types of soil at the project site. To know what exactly we deal with, we are required to classify each soil and determine its characteristics such as grain-size distribution and plasticity. This knowledge will allow us to select the most appropriate design method. The following section will explain how to classify coarse-grained and fine-grained soils using data from laboratory tests.

4.1 Size of soil fractions

Soil consists of particles that have different sizes; it may include gravel, sand, silt and/or clay. The size of each fraction is defined by the testing standard which we follow to perform particle size distribution analysis. As a few standards exist (for example, ASTM [USA], British Standard [BS] and Australian Standard [AS] in Figure 4.1), we should be aware of some minor difference in the fraction boundaries.

Clay	Silt	Sand	Gravel	Particle size, mm
0.002	0.075	4.75	75	USCS
0.002	0.060	2.00	60	BS
0.002	0.075	2.36	63	AS

Figure 4.1 Boundaries between the soil fractions used in different standards (USCS-Unified Soil Classification System, BS-British Standard, AS-Australian Standard).

Question: *What standard shall we use?*
Answer: You should follow the standard used in your country. However, as the difference between the standards seems to be rather negligible (for example, 0.075 mm [USCS] vs. 0.063 mm [BS]), the final outcome is often very similar regardless of the standard used.

4.2 Laboratory work: sieve test and analysis

The amount of each fraction can be found by means of relatively simple sieve tests and, when necessary, hydrometer tests. The sieve test is commonly used in practice and numerous videos of this laboratory experiment are freely available on the Internet. In this test, an

Table 4.1 Results of sieve test.

Soil preparation	Oven-dried soil (dried at 105°C)
Total sample mass before sieving (g)	879.42
Total sample mass after sieving (g)	875.35
Percent soil loss during sieving (%)	0.5

Table 4.2 Results of sieve test.

Sieve opening (mm)	Sieve mass (g)	Mass of sieve + soil (g)	Mass of soil retained (g)	Cumulative soil mass retained (g)	Cumulative percent retained %	Percent passing %
(1)	(2)	(3)	(4)	(5)	(6)	(7)
			Column (4) = Column (3) − Column (2)	Column (5) = Column (4) + Column (5) of the line above	Column (6) = (Column (5) ÷ Total mass of soil) × 100%	Column (7) = 100% − Column (6)
37.5				0		
19	687.45	693.33	5.88	5.88	0.67	99.3
13.2	628.84	696.87	68.03	73.91	8.44	91.6
9.5	610.34	711.52	101.18	175.09	20.00	80.0
4.75	549.63	777.21	227.58	402.67	46.00	54.0
2.36	535.62	730.28	194.66	597.33	68.24	31.8
1.18	504.15	628.55	124.4	721.73	82.45	17.5
0.6	469.69	541.84	72.15	793.88	90.69	9.3
0.425	448.15	470.20	22.05	815.93	93.21	6.8
0.3	434.23	450.97	16.74	832.67	95.12	4.9
0.15	408.94	431.67	22.73	855.4	97.72	2.3
0.075	408.84	420.99	12.15	867.55	99.11	0.9
Pan	240.49	248.29	7.8	875.35	100.00	0.0

oven-dried soil sample is sieved through a stack of sieves. Each sieve has a mesh of a certain size so that it does not allow particles smaller than the hole in the mesh to pass through it. The sieves are arranged in order where the largest one is on top and the finest sieve is at the bottom. A pan is placed below the bottom sieve to collect the soil that passes the finest sieve. By measuring the mass of soil retained in each mesh and expressing it as a percentage of the total, the grain size distribution curve is obtained.

Results from a sieve test on coarse-grained soil are given in Tables 4.1 and 4.2 (Columns 1–3). We will interpret and analyze the data by completing Table 4.2 (Columns 4–7).

Question: Why is the total mass after sieving smaller than the original mass?
Answer: During this test, soil may be lost by escaping out the sides of sieves, or becoming lodged in the meshes of the sieves. It is known as the sieve loss.

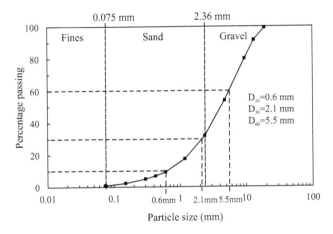

Figure 4.2 A particle size distribution curve from the sieve test (the data used to plot this graph are taken from Table 4.2, Columns 1 and 7).

To develop a particle size distribution curve of this soil (Fig. 4.2), we will plot the experimental data from Columns 1 and 7 of Table 4.2.

Analysis of the particle size distribution curve (Fig. 4.2) shows that about 32% of the soil is finer (passing) than 2.36 mm, indicating that 68% ($100 - 32 = 68$) of the soil particles are larger than 2.36 mm, i.e., gravel. The fines content (particles finer than 0.075 mm) is about 1% while the amount of sand is 31% ($32 - 1 = 31$). Note that this soil contains significant amount of gravel, and it will later be classified as "gravel" (see details in Section 4.5).

Question: *Do we need to perform a hydrometer test for this soil?*
Answer: The hydrometer test is only required when the soil contains more than 10% fines (silt + clay >10%). In this example, the fine content is only 1%, indicating that a hydrometer test is not necessary.

Question: *Can we use the sieve test to determine the amount of silt and clay instead of hydrometer tests?*
Answer: No. In hydrometer tests, only fines (soil material less than 0.075 mm) are used to determine the percentage of silt and clay. The sieve test does not work for fine material because the size of fine particles is too small, which makes it extremely difficult to manufacture sieves with the meshes finer than 0.075 mm. Also, smaller clay particles tend to form clods and cannot pass through the meshes individually.

4.3 Soil gradation

Coarse-grained soil is commonly classified based on its gradation. It can be described as "well-graded" or "poorly graded". The latter includes the "uniform" (Fig. 4.3b) and "gap-graded" (Fig. 4.3c) soil structures. Well-graded soils (Fig. 4.3a) contain a wide range of soil particles and they typically have greater densities than poorly graded soils. Poorly graded soils do not have all particle sizes (uniform) or can have at least one particle size missing (gap-graded).

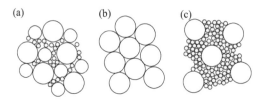

Figure 4.3 Well-, poorly and gap-graded soils.

To determine soil gradation, it is necessary to obtain the coefficients of curvature (C_c) and uniformity (C_u), which are related to certain points on the gradation curve (Equations 4.1 and 4.2).

$$C_c = \frac{D_{30}^{\,2}}{D_{10}D_{60}} \tag{4.1}$$

$$C_u = \frac{D_{60}}{D_{10}} \tag{4.2}$$

where D_{10}, D_{30} and D_{60} are referred to the particle sizes that are finer than 10%, 30% and 60%, respectively. For coarse-grained soil to be well-graded, the following criteria for gravel ($C_u > 4$, $1 < C_c < 3$) and for sand ($C_u > 6$ and $1 < C_c < 3$) must be satisfied. Otherwise, the soil is considered to be poorly graded.

For the soil in Figure 4.2, we can obtain

$D_{10} \approx 0.6$ mm, $D_{30} \approx 2.1$ mm, $D_{60} \approx 5.5$ mm

We will calculate the coefficient of uniformity as

$$C_u = \frac{D_{60}}{D_{10}} = \frac{5.5}{0.6} \approx 9.2$$

The coefficient of curvature, $C_c = \dfrac{D_{30}^{\,2}}{D_{10} \cdot D_{60}} = \dfrac{2.1^2}{0.6 \cdot 5.5} \approx 1.34$

This soil is well-graded because both C_u and C_c meet the criteria for gravel ($C_u > 4$, $1 < C_c < 3$).

Question: *Is it necessary to determine the gradation of coarse-grained material every time we perform sieve analysis?*
Answer: Yes, because we classify coarse-grained soil based on its gradation. Also, it is important to know soil gradation for many practical cases related to soil filters where material of certain gradation is required.

4.4 Clay fraction, clay minerals and clay properties

Particles less than 0.002 mm in size are called clay particles. Clay particles have a strong effect on the soil properties such as plasticity, strength and compressibility. Clay particles are made of clay minerals, among which the most common are montmorillonite or smectite, kaolinite and illite. Common non-clay minerals, which are also present in fine-grained soil, are quartz, feldspar and calcite.

4.4.1 Atterberg limits

Fine-grained soils occur in nature in different states (consistency). Dry fine-grained soils can be relatively hard; however, when saturated with water, they may become extremely weak. To describe the state of fine-grained soils, plastic and liquid limits are used (Fig. 4.4). The plastic limit (PL) is the water content at which the soil changes its state from semisolid to plastic. The liquid limit (LL) refers to the water content exceeding which the soil behaves more like water. To determine the plastic and liquid limit, Atterberg limits tests are performed. Plasticity index (PI) is the difference between the liquid and plastic limits (PI=LL-PL) and it can serve as an indicator of the swell potential of clays. Certain clay minerals including montmorillonite and smectite have the ability to attract and hold water molecules to their surface. This results in higher values of LL and PL.

Question: *What are common values of LL and PL?*
Answer: It can vary a lot, depending on the clay content and clay mineralogy. For soils with kaolinite, LL can be around 35–40%, sometimes reaching 95%. PL typically varies between 25–35%. For soils with montmorillonite and smectite, the liquid limit can easily reach 500–700% (bentonite soil) while PL would typically be in the range of 35–45% (Gratchev et al., 2006; Gratchev and Sassa, 2015).

When we dry saturated soil, its volume will gradually decrease until it becomes constant. The water content at which the soil volume does not change any more is defined as the shrinkage limit (Fig. 4.4).

Question: *Can we use the same approach of soil gradation (poorly or well-graded) for plastic soils?*
Answer: It does not work well for plastic soils because the properties of such deposits strongly depend on soil plasticity. There are empirical correlations between soil plasticity and other properties (including strength) that can be considered for design purposes.

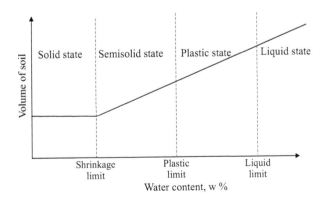

Figure 4.4 Definitions of plastic, liquid and shrinkage limits.

4.4.2 Laboratory work: Atterberg limits tests and analysis

Liquid limit is defined as the water content at which the soil starts to act as a liquid. The test procedure involves mixing soil with various amount of water and determining its consistency using a liquid limit device. In this test, moist soil is placed in the cup of the liquid limit device and a groove is cut on the middle of the soil sample. By turning the crank on the device, the cup will drop from a height of 1 cm and the soil will move towards closing the groove. The number of cranks (blows) necessary to close the groove over a length of about 1.3 cm (0.5 in.) is recorded (see Table 4.3, Column 1) as well as the moisture content of the soil (see Table 4.3) in the cup.

The procedure is repeated (at least 3 times) to obtain a range of blow counts (from 10 to 40), and analysis is performed as shown in Figure 4.5, where the liquid limit is determined as the moisture content corresponding to 25 cranks (blow counts).

The plastic limit is defined as the moisture content at which a 0.3-cm (0.125 in) diameter rod of soil begins to crumble. The procedure is repeated three times as shown in Table 4.4 and an average value of PL is reported.

Table 4.3 Results of liquid limit tests.

Number of cranks (Blow count)	Container ID	Mass of container (M_c)	Mass of wet soil + container (M_1)	Mass of dry soil + container (M_2)	Moisture content (w), %
33	La1	9.12	20.16	16.73	45.1
24	La2	8.69	18.73	15.49	47.6
15	La3	8.85	20.09	16.35	49.9
10	La4	21.59	32.09	28.55	50.9
6	La5	22.44	36.44	31.60	52.8

Note that $w = \dfrac{(M_1 - M_2)}{(M_2 - M_c)} \cdot 100\%$

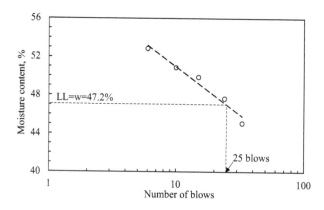

Figure 4.5 Analysis of laboratory data from five liquid limit tests. Note that the water content is plotted against the logarithm of the number of blows.

Table 4.4 Results of plastic limit tests.

Container ID	Mass of container (M_c)	Mass of wet soil + container (M_1)	Mass of dry soil + container (M_2)	Moisture content (w), %
Pb1	23.66	26.13	25.69	21.7
Pb2	9.58	13.35	12.61	24.2
Pb3	9.33	11.35	10.98	22.4
Average Plastic Limit (PL)				23.0
Corresponding Liquid Limit (LL) (From Figure 4.5)				47.2
Plasticity Index (PI = LL-PL)				24.2

Note that $w = \dfrac{(M_1 - M_2)}{(M_2 - M_c)} 100\%$

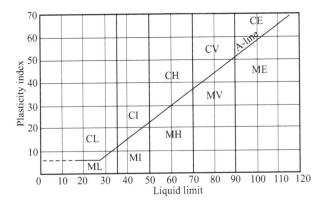

Figure 4.6 Plasticity chart. A-line, which is the boundary between clay (C) and silt (M), is defined as PI = 0.73 (LL-20).

4.4.3 The plasticity chart

For fine-grained soils, plasticity appears to be the most important property as it governs the soil behavior. For this reason, such soils are classified based on their plasticity using the plasticity chart (Fig. 4.6). In this chart, the A-line (described as PI = 0.73[LL-20]) defines the type of soil. All soils above the A-line (PI > 0.73[LL-20]) are classified as "clay" (symbol C) while soils below the A-line (PI < 0.73[LL-20]) are considered to be "silt" (symbol M). Other symbols are: L – low plasticity (lean), I – intermediate plasticity, H – high plasticity (fat), V – very high plasticity and E – extremely high plasticity.

4.5 Soil classification

Soils that have a significant amount of coarse material (gravel and sand) are referred to as "coarse-grained" soils. According to the AASHTO (American Association of State Highway and Transportation Officials) classification system, coarse-grained soils contain less than 35% fines while for the Unified Soil Classification System (USCS), the boundary between coarse-grained and fine-grained soils is 50%. Due to small fines content, coarse-grained soils generally do not exhibit plasticity and, in such a case, they are classified according to the percentage of each fraction and gradation using either the USCS chart (Fig. 4.7) or AASHTO flow chart (Fig. 4.8).

Question: *What is the difference between these two classification systems?*
Answer: Both systems use a similar concept where the soil is classified based on the percentage of each fraction, gradation and plasticity. These classification systems commonly produce similar results. For example, for the soil whose particle size analysis was given in Table 4.2, the classification symbol will be GW, according to both classification systems. The major difference is the division between coarse-grained and fine-grained soils: for USCS the boundary is at 50% fines while for the AASHTO classification, it is 35% fines.

When soil is plastic (even coarse-grained soil can be plastic when it has sufficient clay content), it is classified based on its plasticity by means of the plasticity chart. For example, if soil has more than 50% fines, LL = 55%, PL = 25% and PI = LL-PL = 30%, it is classified as "fat clay" (USCS symbol is "CH"), according to the USCS classification systems (Fig. 4.7).

% passing #200 (0.075 mm)	% passing #4 (4.75 mm)	% passing #200 (0.075 mm)			USCS Symbol	USCS Name
<50%	>50%	0-5%	C_u>6 and 1<C_c<3?	yes	SW	Well-graded sand
				no	SP	Poorly-graded sand
		5-12%	Dual classification		SP-SM	Poorly-graded sand with silt
					SP-SC	Poorly-graded sand with clay
					SW-SM	Well-graded sand with silt
					SW-SC	Well-graded sand with clay
		12-50%	PI>0.73(LL-20) %?	yes	SC	Clayey sand
				no	SM	Silty sand
	<50%	0-5%	C_u>4 and 1<C_c<3?	yes	GW	Well-graded gravel
				no	GP	Poorly-graded gravel
		5-12%	Dual classification		GP-GM	Poorly-graded gravel with silt
					GP-GC	Poorly-graded gravel with clay
					GW-GM	Well-graded gravel with silt
					GW-GC	Well-graded gravel with clay
		12-50%	PI>0.73(LL-20) %?	yes	GC	Clayey gravel
				no	GM	Silty gravel

% passing #200 (0.075 mm)	LL>50%?	PI>0.73(LL-20) %?	USCS Symbol	USCS Name
>50%	yes	yes	CH	Fat clay
		no	MH	Elastic clay
	no	yes	CL	Lean clay
		no	ML	Lean silt

Figure 4.7 Unified Soil Classification System (USCS) chart.

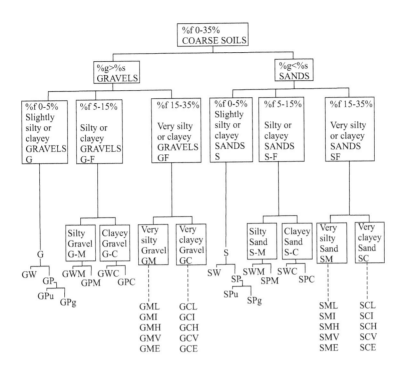

Figure 4.8a AASHTO classification chart for coarse-grained soil.

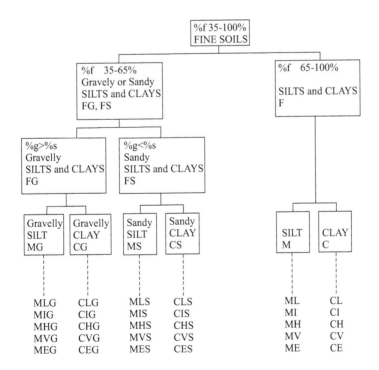

Figure 4.8b AASHTO classification chart for fine-grained soil.

4.6 Clay activity and liquidity index

Plastic fine-grained soils can also be classified based on their activity (Table 4.5) which is the ratio of the plasticity index (PI) to the percentage of clay fraction (Equation 4.3). The more active the fine-grained soil is, the more problems it will likely cause during construction.

$$Activity = \frac{PI}{\%clay} \tag{4.3}$$

The *Liquidity index* (LI) indicates the state (consistency) of fine-grained soils. It is related to the soil liquid (LL) and plastic (PL) limits as well as *in-situ* water content (w) (Equation 4.4). Soils with LI > 1 will likely cause significant problems during construction.

$$LI = \frac{w - PL}{PI} = \frac{w - PL}{LL - PL} \tag{4.4}$$

Based on LI, soils can be classified as follows:

LI < 0 (A), brittle fracture if sheared
0 < LI < 1 (B), plastic solid if sheared
LI > 1 (C), viscous liquid if sheared

Question: *Why do we need to know these indices? Isn't it enough just to classify soil using either the USCS or AASHTO classification systems?*
Answer: In some cases, we would like to know more about clay than the standard classifications could give. The indices like "Activity" and "Liquidity index" tend to provide more detailed information about the soil properties and *in-situ* state.

Question: *Can you give any examples of problematic soils?*
Answer: Yes, two examples are expansive/reactive soils and quick clays. The former can swell a lot and cause significant damage to light engineering structures. Reactive soils are commonly identified by their high liquid limit (>50%) and linear shrinkage (more than 8%). Quick clays are soil that can lose a large portion of its strength when it is disturbed. They are commonly found in Scandinavian countries and Canada where they have caused significant geotechnical problems including landslides (a video of the quick clay landslide at Rissa that occurred in 1978 can be easily found on the Internet with detailed explanation of this phenomenon). The sensitivity index (S_t, Equation 4.5) is used to identify the quick clay: when S_t is in the range of 2–4, the soil is considered "low sensitive"; however, as S_t increases, the soil

Table 4.5 Soil classification based on activity.

Description	Activity
Inactive clays	<0.75
Normal clays	0.75–1.25
Active clays	1.25–2
Highly active clays	>2

sensitivity also increases until it becomes "quick" clay ($S_t = 16$), according to the US classification standard (Holtz and Kavocs, 1981).

$$S_t = \frac{Soil\ strength\left(undisturbed\right)}{Soil\ strength\left(disturbed\right)} \tag{4.5}$$

4.7 Project analysis: soil classification

Using the laboratory data given in Tables 1.2–1.4, we will draw particle size distribution curves of two plastic soils (alluvium and residual soil) and non-plastic sand, and classify them according to the USCS and AASHTO classification systems.

For the *sand* soil, we will have the following percentage of each fraction:

Gravel (G) = 100 – 98 = 2%, Sand (S) = 98 – 3 = 95%, Fines (silt + clay) = 3%,

To determine the soil gradation, we will obtain the following values from the curve:

$D_{10} \approx 0.12$ mm, $D_{30} \approx 0.23$ mm, $D_{60} \approx 0.12$ mm,

Using Equations 4.1 and 4.2, we will calculate

$C_c \approx 1.16$, and $C_u \approx 3.17$

According to USCS and AASHTO classification systems, this soil is classified as poorly graded sand (SP).

For the *residual* soil, the amount of each fraction will be

G = 100 – 75 = 25%, S = 75 – 35 = 40%, Silt (M) = 35 – 12 = 23%, Clay (C) = 12%.

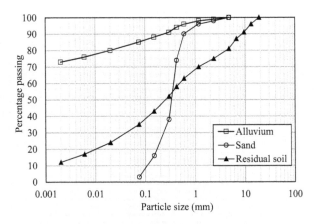

Figure 4.9 Grain size distribution curves.

Using the plasticity chart (Fig. 4.6) and the soil plasticity characteristics (LL = 29%, PI = 11%) from Table 1.3, we will classify this soil as "CL".

According to USCS classification system, this soil is clayey sand (SC) while it is classified as "SCL" following the AASHTO system.

Question: *For the residual soil, the amount of fines is 35%, putting it on the boundary between coarse-grained and fine-grained soils in the AASHTO classification system. Is it coarse-grained or fine-grained soil?*

Answer: If the amount is 35%, we still consider this soil as coarse-grained. When it is more than 35%, it becomes fine-grained soil.

Silty clay (Alluvium):

$$G = 100 - 99 = 1\%, \ S = 99 - 85 = 14\%, \ M = 85 - 73 = 12\%, \ C = 73\%$$

Using the plasticity chart (Fig. 4.6) and LL=67%, PI=38% from Table 1.2, the soil is classified as "CH". According to USCS and AASHTO classification systems, this soil is classified as CH.

The classification based on liquidity index and activity will give:

$$LI = \frac{w - PL}{PI} = \frac{78.3 - 29}{38} \approx 1.3$$

This soil is classified as "viscous liquid when shearing"

$$A = \frac{PI}{\%clay} = \frac{38}{73} \approx 0.52$$

This soil is classified as "inactive clay"

All obtained results are summarized in Table 4.6.

Table 4.6 Classification of soils from the project site.

	Silty Clay (Alluvium)	Sand (Pit 1)	Sandy Silty Clay (Residual soil)
Depth (m)	2.6–3.0 m	8.2–8.6 m	10.1–10.5 m
Plasticity index	67 – 29 = 38	N/A	29 – 18 = 11
Gravel (G), %	1	2	25
Sand (S), %	14	95	40
Silt (M), %	12	3	23
Clay (C), %	73		12
Classification: (AASHTO symbols)	CH	SP	SCL
Classification: (USCS symbols)	CH	SP	SC
Liquidity index + Classification based on Liquidity index	1.3 Viscous	N/A	0.1 Plastic
Activity + Classification based on Activity	0.52 Inactive	N/A	0.92 Normal

4.8 Problems for practice

Problem 4.1 Particle size distribution curves for Soils A and B are given in Figure 4.10. For Soil B, Liquid limit = 48%, Plastic limit = 34%.

Using the AASHTO soil classification system and the plasticity chart, classify Soils A and B.

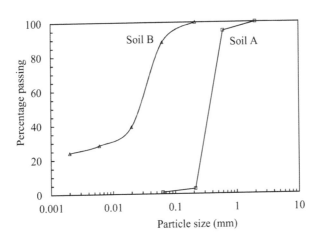

Figure 4.10 Particle size distribution curves of Soils A and B.

Solution

In this example, we will follow the British Standard to determine the amount of each soil fraction. Soil A is non-plastic soil; for this reason, we will use gradation to classify it.

Gravel = 0%, Sand = 98.5%, Fines = 1.5%

To determine the soil gradation, we will obtain:

$D_{10} \approx 0.24$ mm, $D_{30} \approx 0.29$ mm, $D_{60} \approx 0.4$ mm

Resulting in

$$C_u = \frac{D_{60}}{D_{10}} = \frac{0.4}{0.24} \approx 1.66 \ , \ C_c = \frac{D_{30}^{2}}{D_{10} \cdot D_{60}} = \frac{0.29^2}{0.4 \cdot 0.24} \approx 0.876 \ .$$

According to AASHTO classification system, it is sand. C_u does not meet the requirement for well-graded sand, indicating that this soil is poorly graded sand (SP).

Question: *Do we know if this soil is "uniform" or "gap-graded"?*
Answer: Yes, this soil has sand as the dominant type (more than 98%), which makes it uniform. The final answer can be SP_u.

Because Soil B is plastic, we are required to use the plasticity chart (Figure 4.6) as well. First, we will find the percentage of each fraction:

Gravel = 0%, Sand = 12%, Silt = 64%, Clay = 12%.

Plasticity index of soil B is

$$PI = LL - PL = 48 - 34 = 12 \%$$

Using the plasticity chart (Figure 4.6), we will classify this soil as MI. According to the AASHTO classification system, it is also MI.

Problem 4.2 Data from laboratory tests on Soils 1 and 2 are given in Table 4.7. Classify Soil 1 and Soil 2.

Table 4.7 Data from sieve, hydrometer and Atterberg limits tests.

Grain size, mm	Soil 1	Soil 2
	% Passing	
19	100	
13.2	97	
9.5	93	
6.7	91	
4.75	87	100
2.36	80	99
1.18	71	98
0.6	52	96
0.425	37	91
0.3	27	77
0.15	18	42
0.075	9	23
0.02		18
0.006		14
0.002		10
Liquid Limit (%)		33
Plastic Limit (%)		21

Solution

In this example, we will classify the soils without drawing their grain size distribution curves. We will follow the Australian standard (AS) which defines the boundaries as 2.36 mm (between gravel and sand), 0.075 mm (between sand and silt) and <0.002 mm (for clay).

Soil 1 is coarse-grained material with fines (F) of 9%. The gravel fraction is G = 100 − 80 = 20% and Sand is S = 80 − 9 = 71%.

Soil gradation analysis gives

$D_{10} \approx 0.085$ mm, $D_{30} \approx 0.32$ mm, $D_{60} \approx 0.8$ mm

Using Equations 4.1 and 4.2, we will get

$C_c \approx 1.5$ and $C_u \approx 94$ (Note that C_c and C_u both meet the requirement for well-graded soil).

According to the AASHTO classification system, this soil can be SWM or SWC.

Question: Why are there two answers?
Answer: We don't know how much silt or clay is in this soil as no data about soil plasticity is provided. For this reason, we give two possible answers.

For Soil 2, the fraction percentage is

$G = 100 - 99 = 1\%$, S $= 99 - 23 = 76\%$, M (Silt) $= 23 - 10 = 13\%$, C $= 10\%$.

Plasticity index is

PI $= 33 - 21 = 12\%$.

It is "CL" according to the plasticity chart (Fig. 4.6) and SCL according to the AASHTO classification system.

4.9 Review quiz

1. Granular soils are classified "loose" or "dense" based on their

   a) porosity   b) relative density
   c) plasticity   d) dry density

2. Which of the following clay minerals is typically associated with the largest liquid limit?

   a) kaolinite   b) illite   c) chlorite   d) montmorillonite

3. Which of the following terms is not used with fine-grained soils?

   a) relative density   b) activity
   c) liquidity index   d) plasticity

4. Examination of a soil sample produced the following results: Gravel $= 41\%$, Sand $=$ 25%, Silt $= 27\%$, Clay $= 7\%$, LL $= 26\%$, PL $= 17\%$, natural water content $= 20\%$. What is the activity of this soil?

   a)   0.7   b)   1.0   c)   1.3   d)   2.1

For Questions 5–7, use Figure 4.11

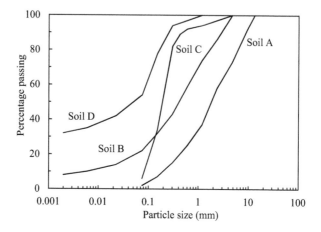

Figure 4.11 Grain size distribution curves.

5. What soil has the largest clay fraction?

   a) A   b) B   c) C   d) D

6. What soil has the largest amount of gravel?

   a) A   b) B   c) C   d) D

7. What soil can be classified as uniform?

   a) A   b) B   c) C   d) D

8. According to the AASHTO Soil Classification System, fine-grained soils are defined as soil with fines content equal to or greater than

   a) 35%   b) 50%   c) 55%   d) 75%

9. What clay has the largest liquid limit?

   a) CE   b) CH   c) CI   d) CV

10. If the liquidity index (LI) is 0, the natural water content of plastic soil is equal to its

    a) liquid limit   b) plastic limit
    c) plasticity index   d) shrinkage limit

Answers: 1) b   2) d   3) a   4) c   5) d   6) a   7) c   8) a   9) a   10) b

Chapter 5

Soil compaction

Project relevance: For construction of roads and embankments, soil material must be compacted to its densest state, which will result in higher soil strength and long-term stability. To obtain the compaction characteristics of soil, engineers conduct a series of laboratory standard compaction tests and then use these data for field compaction. In this project, an embankment will be built as part of a pre-load method to consolidate the soft alluvial clay (more details about consolidation are given in Chapter 10) using the sand from Pit 1. We are required to interpret and analyse data from compaction tests on this sand. This section will introduce the concept of soil compaction and discuss how to determine compaction characteristics of soil.

5.1 Compaction process

Compaction is the process of packing soil particles by mechanical means to decrease the air content in soil, and thus increasing the soil density. It is used for road and dam constructions where soils are required to be compacted to their maximum dry density (i.e., densest state). At this state, the soil exhibits the most favorable geotechnical properties such as high density and high strength. Compacted soils also provide high CBR (California Bearing Ratio) values which are necessary for pavement and railway design (Gratchev et al., 2018).

To determine compaction characteristics of soil, a series of standard Proctor compaction tests are performed on soil specimens with different water content. Results of such tests are plotted as the water content against the dry density as shown in Figure 5.1, and values of the *maximum dry density* (ρ_{dmax}) and the corresponding water content (*optimum water content*, w_{opt}) are obtained from the graph.

Question: *Why do we use the dry density instead of bulk density? It seems easier to obtain the bulk density from field measurements while the dry density requires extra knowledge of water content?*

Answer: We should use two independent parameters (which are dry density and water content) to obtain a compaction curve. The bulk density is not suitable for this purpose because it is directly related to the water content; i.e., when the water content increases the bulk density also increases.

5.2 Laboratory work: compaction tests and analysis

To develop a compaction curve, at least five soil specimens should be compacted. For each test, the water content should increase in a way that at least two specimens are dry of w_{opt} and two specimens are wet of w_{opt}. The test procedure requires a standard proctor compaction hammer and mould. The moist specimen is compacted in the mould in three layers, and for each layer, 25 hammer blows/lift are used. The mass of the compacted soil is recorded at the end of each test, and the water content of the top and bottom part of the soil specimen is measured. Results of five compaction tests are given in Table 5.2 while water content measurements are summarized in Table 5.3.

Table 5.1 Compaction test details.

Compaction effort: Standard	Mould diameter: 105 mm
Soil hydration period prior compaction: 24 h	Mould volume (V_m): 1000 cm³
Mould height: 115.4 mm	Mass of mould (m_1): 4558 g

Table 5.2 Data from compaction tests.

Test No.	1	2	3	4	5
Mass of compacted soil + mould, m_2 (g)	6192	6293	6436	6555	6460
Mass of soil (g), $= (m_2 - m_1)$	1634	1735	1878	1997	1902
Wet density (g cm⁻³), $\rho = (m_2 - m_1)/V_m$	1.635	1.736	1.879	1.998	1.903
Water content w (%) from Table 5.3	6.9	11.6	15.1	17.9	23.6
Dry density (g cm⁻³), $\rho_d = \rho/(1 + w)$	1.53	1.56	1.63	1.70	1.54

Table 5.3 Data on water content.

Test No.	1		2		3		4		5	
Location within specimen	Top	Bottom	Top	Bottom	Top	Bottom	Top	Bottom	Top	Bottom
Container ID	T1	B1	T2	B2	T3	B3	T4	B4	T5	B5
Mass of container, m_c (g)	8.98	9.34	21.79	9.52	20.93	21.6	21.53	9.92	33.71	22.26
Mass of container and wet soil, m_w (g)	26.37	24.42	33.67	24.99	34.62	29.59	33.05	29.89	66.53	63.59
Mass of container and dry soil, m_d (g)	25.26	23.44	32.44	23.37	32.74	28.59	31.42	26.67	60.57	55.33
Mass of soil solids (g), $M_s = m_d - m_c$	16.28	14.1	10.65	13.85	11.81	6.99	9.89	16.75	26.86	33.07
Mass of water (g), $M_w = m_w - m_d$	1.11	0.98	1.23	1.62	1.88	1.00	1.63	3.22	5.96	8.26
Water content, $w\%, = (M_w/M_s) \times 100\%$	6.81	6.95	11.54	11.69	15.91	14.30	16.48	19.22	22.18	24.97
Average water content, %	6.9		11.6		15.1		17.9		23.6	

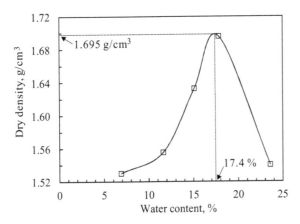

Figure 5.1 A standard compaction curve. Maximum dry density (ρ_{dmax}) is about 1.70 g cm^{-3} and the optimum water content (w_{opt}) is 17.4%.

When the dry density and corresponding water content are obtained for each test, the data are plotted as shown in Figure 5.1 to determine the maximum dry density ($\rho_{dmax} \approx 1.70\ \mathrm{g\ cm^{-3}}$) and optimum water content ($w_{opt} \approx 17.4\%$).

Question: *Is there any difference between compaction characteristics of coarse-grained and fine-grained soils?*
Answer: Compaction curves of coarse-grained soils generally exhibit more pronounced peaks compared to fine-grained soils. In addition, coarse-grained soils tend to have higher values of ρ_{dmax}. In contrast, fine-grained soils require more moisture for better compaction, and for this reason, they typically have greater values of w_{opt} (range of 15–20%) compared to coarse-grained soils (10–15%) (Gratchev et al., 2018).

5.3 Compaction in the field

The values of the maximum dry density and optimum water content are necessary to perform compaction work in the field. Different equipment, which is briefly described in Table 5.4, can be used at construction sites depending on the soil type and the thickness of soil layer.

Table 5.4 Type of construction in practice.

Equipment	Purpose
Smooth wheeled roller	Effective for coarse-grained soils
Sheepsfoot roller	Commonly used for fine-grained soils
Impact roller	Required for deeper (2–3m) compaction
Dynamic compaction using a pounder (tamper)	When a tamper is dropped from certain heights, deep compaction can be achieved
Dynamic compaction using Vibroflot	Effective mostly for granular material, providing a higher level of compaction

Question: *It seems very difficult to compact soil to its maximum dry density in the field, especially considering the procedures and equipment used for this purpose. What is the quality control for such work?*

Answer: Yes, it is not always possible to compact soil to its maximum dry density. For this reason, an acceptable range of dry density (95% or more of the maximum dry density) is used to control the field compaction work. A *sand cone test* is commonly carried out at the compaction site to determine the dry density and water content of the compacted soil. To ensure a high level of compaction, it is required to perform at least one test per 1000 m³ of compacted soil.

5.4 Project analysis: soil compaction

We have already established that the alluvial soft clay is too loose and weak, and its properties need to be improved prior to construction. One of the options is a pre-load method where sand fill (embankment) is placed on top of the ground. This will generate an additional load on the alluvial clay layer, initiating some settlements (it is called "consolidation"; we will deal with this process in Chapter 10). It is economical to use the sand material from Pit 1 as it is very close to the site; however, this sand needs to be compacted to its maximum dry density. To determine this parameter, five standard Proctor compaction tests were performed and their results are summarized in Table 5.5. Note that the volume of the compaction mould is 1000 cm³.

Table 5.5 Results from a series of standard compaction tests.

Mass of compacted soil (g)	1879	1982	2091	2123	2105	2075
Moisture content (%)	12	14	16	18	20	22

Solution

First, we will calculate the bulk density (Equation 3.5) and dry density (Equation 3.10) for each test (Table 5.6), and then we will plot the data as the dry density against water content to obtain a standard compaction curve (Fig. 5.2).

From Figure 5.2, the maximum dry density (ρ_{dmax}) is about 1.81 g cm^{-3} and the optimum water content (w_{opt}) is 16.9%. The value of 1.72 g cm^{-3} makes 95% maximum dry density.

Volume of sand necessary to construct the embankment. It is estimated that *in-situ* bulk density (ρ) of the sand from Pit 1 would be 1.55 g cm^{-3} and water content (w) would be 7% at the time of embankment construction. Calculate the volume (V_{pit}) of the sand required for 1 m³ of embankment (V_{emb}).

Table 5.6 Analysis of data from a series of compaction tests.

Mass of compacted soil (g)	1879	1982	2091	2123	2105	2075
Water Content (%)	12	14	16	18	20	22
Bulk density (g cm^{-3})	1.879	1.982	2.091	2.123	2.105	2.075
Dry density (g cm^{-3})	1.68	1.74	1.80	1.80	1.75	1.70

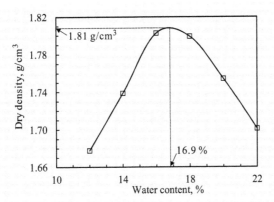

Figure 5.2 Compaction curve of the sand from Pit I.

Solution

We will use the volumes and void ratios of sand in the field (V_{pit}, e_{pit}) and the same sand but compacted to its maximum dry density in the embankment (V_{emb}, e_{emb}) as given in Equation 5.1.

$$\frac{V_{pit}}{V_{emb}} = \frac{\left(1+e_{pit}\right)}{\left(1+e_{emb}\right)} \tag{5.1}$$

Let's find e_{pit} using the dry density and specific gravity ($G_s = 2.52$ from Table 3.3). For the *in-situ* conditions, the dry density will be

$$\rho_d = \frac{\rho}{1+w} = \frac{1.55}{1+0.07} \approx 1.45 \text{g/cm}^3$$

By re-arranging $\rho_d = \dfrac{G_s \rho_w}{1+e}$

We will get

$$e = \frac{G_s \rho_w}{\rho_d} - 1$$

resulting in

$e_{pit} = 0.843$.

This sand compacted in the embankment will have $\rho_{dmax} = 1.81$ g cm^{-3} (Figure 5.2), resulting in $e_{emb} = 0.475$.

Then, we will have

$$\frac{V_{pit}}{V_{emb}} = \frac{\left(1+e_{pit}\right)}{\left(1+e_{emb}\right)} = \frac{V_{pit}}{1} = \frac{\left(1+0.843\right)}{\left(1+0.475\right)}$$

Now we can obtain the volume of sand from the pit required for 1 m³ of embankment as

$$V_{pit} \approx 1.25\, m^3$$

Field work quality control

Let's assume that the embankment was already constructed and two samples were collected for quality control. The following data about the sample mass and volume were recorded (Table 5.7). The compaction specifications require that the dry density of the compacted sand should be greater than or equal to 95% of the maximum dry density (i.e., ≥ 1.72 g cm⁻³) while the water content should be within $\pm 1.5\%$ of the optimum water content (w_{opt} = 16.9%). We are required to establish whether these control tests meet the specifications.

Table 5.7 Data from two compaction control tests.

Control test No.	Volume of soil (cm³)	Mass of wet soil (g)	Mass of dry soil (g)
1	1200	2471	1952
2	950	2012	1721

Solution

We will find the bulk and dry density for each soil sample and compare the results with the specifications (Table 5.8).

Table 5.8 Comparisons of filed test results with the compaction requirements.

Control test No.	Dry density, ρ_d (g cm⁻³)	Water content, w (%)	Comments
1	1.63	26.6	No, because it is too wet
2	1.81	16.9	Yes, it meets the requirements

5.5 Problems for practice

Problem 5.1 The results from a series of standard compaction tests are summarized in Table 5.9. The specific gravity of the solids is 2.64 and the volume of the compaction mould is 1000 cm³.

Table 5.9 Data from standard Proctor compaction tests.

Mass of compacted soil (g)	1555	1782	1905	1902	1862	1820
Water content (%)	8	12	15	18	20	23

a) Draw the compaction curve and determine the optimum water content and maximum dry density.

The compaction specifications require that the *in-situ* dry density be greater than or equal to 95% of the maximum dry density from the Proctor compaction test and for the water content to be within ±1.5% of the optimum water content. Compaction control tests were carried out at two different locations and the results are given in Table 5.10.

Table 5.10 Data from compaction control tests.

Control test No.	Volume of soil (cm³)	Mass of wet soil (g)	Mass of dry soil (g)
1	1304	2215	1882
2	1036	1924	1680

b)   Determine which of the control tests meets the specifications.

Solution

a)   First, we will calculate the density (ρ) of soil for each test using the mass (from Table 5.9) and volume (V = 1000 cm³) of compacted soil. For Test 1, we will get

$$\rho = 1555/1000 \approx 1.56 \text{g cm}^{-3}$$

Then we will find the dry density using the water content (w) as

$$\rho_d = \frac{\rho}{1+w} = \frac{1.56}{1+0.08} \approx 1.44 \text{g cm}^{-3}$$

Once the dry density is obtained for each test, we can draw a standard compaction curve as shown in Figure 5.3 and determine the maximum dry density ($\rho_{dmax} \approx 1.66$ g cm^{-3}) and optimum water content (w $\approx$ 15%) from the graph.

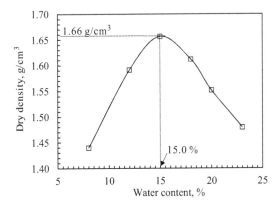

Figure 5.3 Results of compaction tests: the maximum dry density ($\rho_{dmax} \approx 1.66$ g cm^{-3}) and optimum water content (w $\approx$ 15%).

b)   For the control requirements, the dry density in the field should be at least 95% of the maximum dry density, which is 1.58 g cm^{-3}, and the water content must be 15% ± 1.5%. For the control tests, we will have the following results (Table 5.11). In conclusion, only Test 2 meets the requirement.

Table 5.11 Comparisons of filed test results with the compaction requirements.

Control test No.	Dry density, ρ_d (g cm^{-3})	Water content, w (%)	Comments
1	1.44	17.7	No, Does not meet the requirement
2	1.62	14.5	Yes, it meets the requirements

Problem 5.2 A series of standard Proctor compaction tests were performed on soil and the obtained results are presented in Figure 5.4. The volume of the standard mould was 1×10^{-3} m^3 and $G_s = 2.65$.

a) What is the degree of saturation of the sample compacted to the maximum dry density.
b) Estimate the volume of air (in m^3) in the sample compacted to 15% water content.

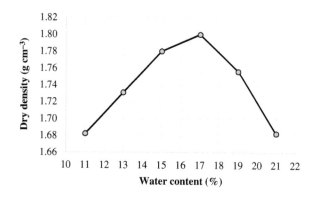

Figure 5.4 Results from standard Proctor compaction tests.

c) The void ratio of this soil in its loosest state is 0.623. Estimate the relative density of the soil sample compacted to 11% water content?

Solution

a) From the graph in Figure 5.4, we will get

$\rho_{dmax} = 1.80$ g cm^{-3}, and w = 17 %.

From $\rho_d = \dfrac{G_s \cdot \rho_w}{1+e}$ we can find the void ratio at the maximum dry density:

$1.8 = \dfrac{2.65 \cdot 1}{1+e} \rightarrow e \approx 0.472$

Using the void ratio, the degree of saturation (S) can be found as

$S = \dfrac{w \cdot G_s}{e} = \dfrac{0.17 \cdot 2.65}{0.472} = 0.954 \text{ or } 95.4 \%$

b) From Figure 5.4, when w = 15%, the dry density is $\rho_d = 1.78$ g cm^{-3}

To find the air content (A_v), we will use the following equation:

$$\rho_d = \frac{G_s \cdot \rho_w (1 - A_v)}{1 + wG_s}$$

We will get

$$1.78 = \frac{2.65 \cdot 1(1 - A_v)}{1 + 0.15 \cdot 2.65} \rightarrow A_v \approx 0.061 \text{ or } 6.1 \%$$

The volume of soil sample, V = 0.001 m³. Using the definition of air content, we can obtain the volume of air (V_a) as

$$V_a = A_v \cdot V \approx 0.000061 \text{m}^3$$

c) In the loosest state, the soil will have the largest void ratio, i.e., $e_{max} = 0.623$

In the densest state, the soil will have the smallest void ratio (e_{min}), which can be achieved at the maximum dry density ($\rho_{dmax} = 1.80$ g cm^{-3}).

By using the following equation, $\rho_{dmax} = \dfrac{G_s \cdot \rho_w}{1 + e_{min}}$

We will find the minimum void ratio, $e_{min} \approx 0.472$

At 15% water content, $\rho_d = \dfrac{G_s \cdot \rho_w}{1 + e} = 1.78 = \dfrac{2.65 \cdot 1}{1 + e}$

We will have $e \approx 0.489$

Relative density, $D_r = \dfrac{\left(e_{max} - e_0\right)}{\left(e_{max} - e_{min}\right)} 100\% = \dfrac{(0.623 - 0.489)}{(0.623 - 0.472)} 100\% \approx 88.7 \%$

According to Table 3.1, this soil is very dense.

5.6 Review quiz

1. Compaction is the process of soil densification that occurs due to the

a) expulsion of organic matter
b) expulsion of air
c) expulsion of water
d) all three process (a-c) are equally important

2. Compaction works best for

a) silt   b) clay
c) gravelly sand   d) all three (a-c)

3. The highest level of compaction for clay can be typically achieved by

    a)   smooth wheeled roller
    b)   sheepsfoot roller
    c)   pounding the ground by a heavy weight
    d)   vibroflotation

4. The zero-air-void curve can be obtained

    a)   through testing in the laboratory conditions
    b)   in the field using the appropriate machinery
    c)   it is a theoretical curve that cannot be achieved in practice

5. How many blows per layer are used in standard Proctor compaction tests?

    a)   15   b)   20   c)   25   d)   30

6. The data from standard Proctor compaction tests are as follows: the maximum dry density of soil is 1.80 g cm^{-3} and the optimum water content is 14%. What will be the minimum value of dry density (in g cm^{-3}) that can still satisfy the quality control requirement in the field?

    a)   1.59   b)   1.71   c)   1.75   d)   1.79

7. In standard Proctor compaction tests, soil is compacted in

    a)   1 layer   b)   2 layers   c)   3 layers   d)   5 layers

8. Increasing compactive effort in compaction tests results in

    a)   greater maximum dry density
    b)   greater optimum water content
    c)   no effect

Answers: 1) b   2) c   3) b   4) c   5) c   6) b   7) c   8) a

Chapter 6

Stresses in soils

Project relevance: During construction, when loads are applied to soil mass, the soft alluvial clay will likely undergo deformations. Large deformations would undermine the stability of soil mass and may pose threats to engineering structures built on this soil. To estimate the amount of deformation, we should first determine the stresses that would act in the soil mass before and after construction. This chapter will introduce different types of stresses that exist in soil and discuss techniques to calculate them.

6.1 Stresses in soil mass

Let's consider a soil element with the area of 1 m² at a depth of z (Fig. 6.1a). There will be total (or normal) stresses (σ_x, σ_y and σ_z) acting on this soil element. In many practical applications, there is no need to carry out a complex three-dimensional stress analysis because most of geotechnical structures including embankments and retaining walls are long in comparisons with their width and height. For this reason, it is often simplified to the stress state (two dimensional) which is shown in Figure 6.1b, where σ_v is the vertical total stress and σ_h is the horizontal total stress.

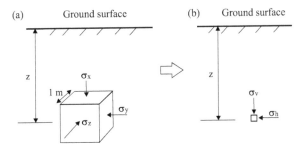

Figure 6.1 A soil element under stresses: a) three-dimensional state of stresses in soil mass; b) simplified stress conditions commonly used in geotechnical applications (σ_v – vertical stress, σ_h-horizontal stress).

Question: *What if we need to perform a 3-D analysis, what shall we do?*
Answer: It is a very difficult task at present but some computer programs are capable of doing this.

The vertical total stress is generated from the weight of the soil mass above the soil element, and it can be obtained using the unit weight (γ) of soil (Equation 6.1).

$$\sigma_v = \gamma z \tag{6.1}$$

6.2 Effective stress and pore water pressure

The total stress consists of the effective stress (σ') and pore water pressure (u). The effective stress is the stress that is being transferred to soil mass through the soil particles (Fig. 6.2). The effective stress determines the properties of soil (including its strength) as it carries the weight of engineering structures. Higher effective stresses result in greater strength of soil and stability of soil mass.

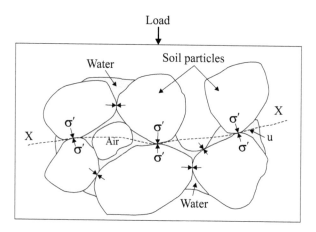

Figure 6.2 Distribution of load in soil mass through soil particles (effective stress, σ') and water (pore water pressure, u) along X-X.

If soil voids are filled with water, the water can also carry the total stress. Such stress is known as the pore water pressure (u). Higher pore water pressures cause geotechnical problems as they tend to decrease the stability of soil.

Question: *The soil strength seems to depend on the effective stress, and if the effective stress decreases, the bearing capacity of soil will also decrease. What will happen when the effective stress drops to zero?*
Answer: It is theoretically possible that the effective stress becomes zero as was shown in some laboratory studies. In the field, the effective stresses can become almost zero when there is a very high increase in the excess pore water pressure. In such a case, the soil loses its strength and behaves more like a liquid. This state is commonly called liquefaction. There are many examples of soil liquefaction triggered by earthquakes or rainfalls that are documented in the literature. Of course, this needs to be avoided in engineering practice as soil liquefaction results in structure collapses or natural disasters like landslides (Gratchev et al., 2006, 2011).

6.2.1 Determination of pore water pressure in the field

The pore water pressure (u) in the field is related to the ground water level (also known as the ground water table), which is measured by means of piezometers (Fig. 6.3). The soil below the ground water level is commonly considered saturated (S = 100%). To calculate the pore water pressure, the distance (h_w) from the point of interest (Point A in Figure 6.3) to the ground water table is measured and used along with the unit weight of water ($\gamma_w = 9.81 \, kN \, m^{-3}$) as shown in Equation 6.2.

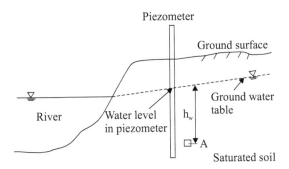

Figure 6.3 A piezometer is used to measure the pressure head (h_w) of water at any point where it is installed.

$$u = \gamma_w h_w \qquad (6.2)$$

Question: What about the soil above the ground water level, is it considered to be dry?
Answer: It is not dry as there is always water that would rise above the ground water table because of the capillary forces. The capillary rise depends on the type of soil: it is low in coarse-grained sand (about 0.1–0.2 m), but it can reach on average of 7–8 m in fine-grained silt and clay. However, in some soil mechanics textbook examples, the soil above the ground water level may be assumed to be dry for simplicity.

6.2.2 Effective stress concept

The effective stress concept is one of the most important concepts in soil mechanics, it relates the normal stress (σ), effective stress (σ') and pore water pressure (u) as shown in Equation 6.3.

$$\sigma' = \sigma - u \qquad (6.3)$$

There are two special cases:

a) *Dry soil* mass (u = 0) when the effective stress is equal to the total stress. It is the best-case scenario as the load from engineering structures is being carried entirely by the soil particles.
b) When the pore water pressure is extremely high and equal to the total stress (u = σ). It is the worst-case scenario as the load from engineering structures (total stress) is being transferred to soil mass only by the pore water, resulting in liquefaction and possible structure damage.

Question: *It seems that when there is lots of rain, the pore water pressure would naturally increase, which may lead to the worst-case scenario. What will be the most effective method in engineering practice to avoid such things?*
Answer: Although it is not possible to avoid rainfalls, properly designed and relatively inexpensive drainage systems would effectively remove extra water from soil, keeping the pore water pressure in soil mass at its minimum.

6.2.3 Horizontal stresses

To calculate the *in-situ* horizontal stress (σ_h) (Fig. 6.1b), the coefficient of earth pressure at rest, K_0, is used (Equation 6.4). The subscript 0 (K_0) is used to indicate the natural, undisturbed earth pressure coefficient in soil mass. It is important to remember that K_0 is defined only for the *effective stress conditions* and it depends on soil type.

$$K_0 = \frac{\sigma_h'}{\sigma_v'} \tag{6.4}$$

In engineering practice, for normally consolidated soils, K_0 is sometimes defined as

$$K_0 = (1 - \sin\phi') \tag{6.5}$$

where ϕ' is the effective friction angle of soil (see Chapter 11 for more details about soil friction angle).

6.3 Excess pore water pressures

6.3.1 Water flow and hydraulic gradient

When water flows in soil, it creates seepage. This may cause unfavorable geotechnical conditions and significantly undermine the stability of soil mass. Let's consider a case where saturated sand is sandwiched between the clay layer and impermeable rock (Fig. 6.4).

Two piezometers installed at Points A and B indicate different water levels, which results in the water flow from Point A to Point B. The rate of water flow depends on hydraulic

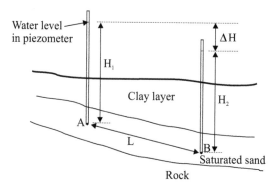

Figure 6.4 Definition of hydraulic gradient.

gradient (i), a parameter which is defined as the ratio between the head difference (ΔH) and the distance (L) between Points A and B (Equation 6.6).

$$i = \frac{\Delta H}{L} \tag{6.6}$$

As can be seen from Equation 6.6, the hydraulic gradient does not depend on soil properties.

6.3.2 Upward seepage

Upward seepage can exist when a layer of aquifer (which is typically coarse-grained soil) is under artesian pressure. For example, the gravel layer in Figure 6.5 is under the artesian pressure, and its pressure head is 10 m above the ground water table. The difference in heads (ΔH) will cause upward seepage through the clay layer, thus generating additional (excess) pore water pressures in the clay layer.

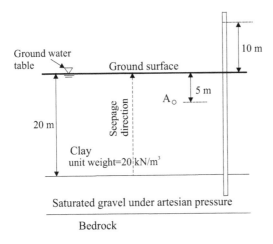

Figure 6.5 Upward seepage in clay.

Question: *What is artesian pressure?*
Answer: Artesian pressure is positive pressure that causes the ground water level to rise. It is named after the former province of Artois in France, where many artesian wells were drilled by Carthusian monks starting from the 12th century.

Let's investigate the effect of upward seepage on the stresses acting at Point A. Point A is located 5 m below the ground surface (z = 5 m), and it is 5 m below the ground water table (h_w = 5 m). The total vertical stress will be

$$\sigma = \gamma \cdot z = 20 \cdot 5 = 100 \, \text{kN m}^{-2}$$

Using Equation 6.2, the pore water pressure (let's call it "hydrostatic" pore water pressure) will be equal to

$$u = \gamma_w \cdot h_w = 9.81 \cdot 5 = 49.1 \, \text{kN m}^{-2}$$

Due to the difference between the ground water table and the level of water in the piezometer ($\Delta H = 10$ m), the hydraulic gradient in the clay layer will be

$$i = \frac{\Delta H}{L} = \frac{10}{20} = 0.5$$

Note that ΔH has its maximum value of 10 m at the bottom of the clay layer (at the point where the upward seepage through the clay layer begins). It gradually decreases ("dissipates") through the clay layer towards the ground water level, where it becomes zero.

To find ΔH_A at Point A, we will re-arrange Equation 6.6 as

$$\Delta H_A = i \cdot L_A$$

The remaining part of the upward water seepage path from Point A to the ground water level (L_A) is 5 m, then

$$\Delta H_A = 0.5 \cdot 5 = 2.5 \, \text{m}$$

By adding this excess pressure head to the already existing ("hydrostatic") head of 5 m, the pore water pressure at Point A will result in

$$u_A = (5 + 2.5) \cdot \gamma_w \approx 73.6 \, \text{kN m}^{-2}$$

6.3.3 Quick conditions

Let's imagine that for some reason, the rate of upward seepage in the soil mass in Figure 6.5 consistently increases, then the value of pore water pressure (u) will also increase, eventually reaching a value of the total stress ($\sigma = u$). In this case, the effective stress drops to zero ($\sigma_v' = 0$), the soil loses its strength and the quick conditions (also known as "boiling" or "liquefaction") occur. This will have a negative effect on the stability of soil mass as well as engineering structures built on it.

The onset of quick conditions is associated with the critical hydraulic gradient (i_c), which is defined using soil specific gravity (G_s) and void ratio (e) (Equation 6.7). For most natural soils, the critical hydraulic gradient is estimated to be approximately 1.

$$i_c = \frac{G_s - 1}{1 + e} = \frac{\gamma_{sat} - \gamma_w}{\gamma_w} \tag{6.7}$$

where γ_{sat} is the unit weight of saturated soil.

Question: *What would be typical examples of liquefaction caused by seepage?*
Answer: A typical example is shown in Figure 6.6, where increases in the river level due to flooding can affect the stability of levees. The higher water level in the river results in greater head differences (see the difference between ΔH and ΔH_f in Figure 6.6), which would add to the existing pore water pressure in soil mass behind the levee, often causing quick conditions (sand boils). There were over 50 failures of the levees and floodwalls protecting New Orleans, Louisiana and its suburbs due to flooding caused Hurricane Katrina in 2005. It was reported that some levee failures were caused by failure of the foundation soils underlying the levees.

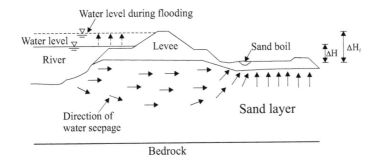

Figure 6.6 Water seepage under a levee caused by flooding.

6.4 Project analysis: stresses and upward seepage

Let's see how upward seepage can affect the stresses at the project site. We will assume that there is sand layer underneath the soft alluvial clay in BH10 as schematically shown in Figure 6.7. This aquifer is under artesian pressure while the ground water table is on the boundary between the topsoil and alluvium. The density of topsoil is 1620 kg m^{-3}, density of the alluvial clay is 1.47 g cm^{-3} and K_0 for the clay layer is 0.9. The water level in a piezometer installed in the sand layer is 1 m above the ground water table.

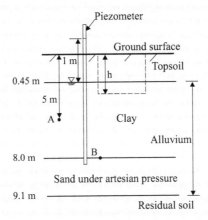

Figure 6.7 Cross-section of study area near BH10.

We will a) calculate the *initial effective vertical stress* and *total horizontal stress* at a depth of 5 m (Point A) before excavation, and b) estimate the maximum cut that can be made in the alluvium so that the stability of this clay layer is not lost. Assume that there are no significant changes in the ground water conditions during the excavation.

Solution

a) Density of topsoil is 1620 kg m^{-3} or 1.62 g cm^{-3}. The unit weight of this soil equals

$$\gamma = 1.62 \cdot 9.81 \approx 15.9 \, \text{kN m}^{-3}$$

The unit weight of the alluvial clay is

$$\gamma = 1.47 \cdot 9.81 \approx 14.4 \, \text{kN m}^{-3}$$

The stresses at Point A are estimated as follows:

The vertical stress is

$$\sigma_v = 0.45 \cdot 15.9 + (5 - 0.45) \cdot 14.4 \approx 72.8 \, \text{kN m}^{-2}$$

Because of the upward seepage in the clay mass, which is caused by the head difference (ΔH) of 1 m, the hydraulic gradient in the clay equals

$$i = \frac{\Delta H}{L} = \frac{1}{8 - 0.45} = 0.132$$

This upward seepage will add to the pore water pressure at Point A, which can be found as the hydrostatic pore water pressure + additional (excess) pore water pressure caused by the upward seepage:

$$u = (5 - 0.45) \cdot 9.81 + \left[0.132 \cdot (5 - 0.45) \cdot 9.81 \right] \approx 50.5 \, \text{kN m}^{-2}$$

The effective vertical stress at Point A will be

$$\sigma_v' = \sigma - u = 72.8 - 50.5 = 22.3 \, \text{kN m}^{-2}$$

The horizontal stresses can be calculated using $K_0 = 0.9$ and Equation 6.4 as

$$\sigma_h' = \sigma_v' \cdot K_0 = 22.3 \cdot 0.9 \approx 20.0 \, \text{kN m}^{-2}$$

The total horizontal stress is

$$\sigma_h = \sigma_h' + u = 70.5 \, \text{kN m}^{-2}$$

b) The excavation of soil may initiate the quick conditions if the cut is too deep and there is no sufficient effective stress ($\sigma_v' \rightarrow 0$) to support the soil stability. The quick conditions would occur at Point B on the boundary of the alluvial clay and sand layers, where the difference in heads is maximum ($\Delta H = 1$ m). Considering this, we will calculate the effective stress at Point B when the cut h is made deep enough to initiate the quick con ditions ($\sigma_v' = 0$).

Normal stress and pore water pressure at Point B are:

$$\sigma = 0.45 \cdot 15.9 + 7.55 \cdot 14.4 \approx 116.0 \, \text{kN m}^{-2}$$
$$u = 7.55 \cdot 9.81 + 1 \cdot 9.81 \approx 83.9 \, \text{kN m}^{-2}$$

The effective normal stress at Point B when the quick conditions would occur is

$$\sigma_v' = \left[116.0 - 0.45 \cdot 15.9 - (h - 0.45) \cdot 14.4 \right] - 83.9 = 0$$

Solving this equation, we will obtain that

$h = 2.18\ m$

Then, the cut in the clay layer would be

2.18–0.45 = 1.73 m

6.5 Problems for practice

Problem 6.1 Soil profile is given in Figure 6.8. The ground water table is at a depth of 1.8 m below the ground surface. For the Silt layer, a soil sample (mass of 26.3 g) was collected from a depth of 2.5 m. It was dried in the oven at 105°C for 24 h and had mass of 22.1 g.

Determine the vertical and horizontal total stresses at a depth of 7.0 m below the ground surface.

Elevation, m

0 ———————————————
        Sand
        w=14%, S=76%, G_s=2.65
1.8 ▽——————————————

        Silt
        G_s=2.70
3.5 ———————————————

        Organic clay
        e=5, ρ_s=2.0 g/cm³, K_o=0.9

7.0 ———————————————
        Clay
        e=1, G_s=2.69, K_o=1.2

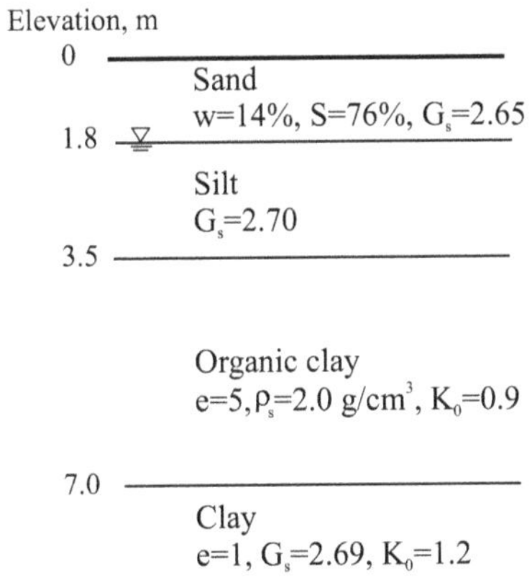

Figure 6.8 Cross-section and soil properties: w is the water content, S is the degree of saturation, G_s is the specific gravity, e is the void ratio, ρ_s is the density of solids, K_o is the coefficient of earth press at rest.

Solution

To calculate the stresses at 7 m, we will first obtain the unit weight for each soil layer above 7 m using the relevant equations from Chapter 3 (Soil constituents).

For the sand layer (Depth: 0 = 1.8 m), we know

w = 14%, S = 76%, G_s = 2.65

We will find

The void ratio, $e = \dfrac{wG_s}{S} = \dfrac{0.14 \cdot 2.65}{0.76} = 0.488$

The unit weight, $\gamma = \dfrac{G_s(1+w)}{1+e}\gamma_w = \dfrac{2.65(1+0.14)}{1+0.488} \cdot 9.81 = 19.9\text{kN m}^{-3}$

For the silt layer (Depth: 1.8–3.5 m), we will find

The water content, $w = \dfrac{26.3 - 22.1}{22.1} = 0.19$

The void ratio, $e = \dfrac{wG_s}{S} = \dfrac{0.19 \cdot 2.7}{1} = 0.513$

The unit weight, $\gamma = \dfrac{G_s + e}{1 + e} \gamma_w = \dfrac{2.7 + 0.513}{1 + 0.513} \cdot 9.81 = 20.8\,\text{kN m}^{-3}$

For the organic clay layer (Depth: 3.5–7.0 m), we will find

The specific gravity, $G_s = \dfrac{\rho_s}{\rho_w} = \dfrac{2}{1} = 2$

The unit weight, $\gamma = \dfrac{G_s + e}{1 + e} \gamma_w = \dfrac{2 + 5}{1 + 5} \cdot 9.81 = 11.4\,\text{kN m}^{-3}$

Normal stress (σ) at 7 m is the sum of stresses originated from each soil layer above 7 m. It is equal to

$$\sigma = 1.8 \cdot 19.9 + 1.7 \cdot 20.8 + 3.5 \cdot 11.4 \approx 111.1\,\text{kN m}^{-2}$$

The pore water pressure (u) at 7 m,

$$u = 5.2 \cdot 9.81 \approx 51.0\,\text{kN m}^{-2}$$

The effective stress (σ') at 7 m,

$$\sigma' = 111.1 - 51.0 = 60.1\,\text{kN m}^{-2}$$

Horizontal stresses can be calculated using the coefficient of earth pressure at rest (K_0). Because the point of interest is on the boundary between two layers; that is, the organic clay and clay, two values of horizontals stresses should be determined: one for the organic clay layer and the other for the clay layer.

For the organic clay layer ($K_0 = 0.9$), the effective and normal stresses are

$$\sigma'_h = \sigma'_v \cdot K_0 = 60.1 \cdot 0.9 \approx 54.1\,\text{kN m}^{-2}$$
$$\sigma_h = \sigma'_h + u = 54.1 + 51 = 105.1\,\text{kN m}^{-2}$$

For the clay layer ($K_0 = 1.2$), the effective and normal stresses are

$$\sigma'_h = 60.1 \cdot 1.2 \approx 72.3\,\text{kN m}^{-2}$$
$$\sigma_h = 72.3 + 51 = 123.3\,\text{kN m}^{-2}$$

Problem 6.2 Saturated clay (thickness of 6 m) has density of 1.7 g cm^{-3}. It is underlain by sand that is under artesian pressure (Fig. 6.9). There is a lake with 4 m of water above the clay.

a) Calculate the effective vertical stress at Point A.
b) Imagine that the water level in the lake decreases due to pumping. At what level of water in the lake will the quick conditions occur at Point B?

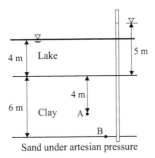

Figure 6.9 Soil profile for Problem 6.2.

Solution

The unit weight of clay, $\gamma = \rho \cdot g = 1.7 \cdot 9.81 = 16.7 \text{kN m}^{-3}$
The total stress at Point A, $\sigma = 4 \cdot 9.81 + 4 \cdot 16.7 = 105.9 \text{kN m}^{-2}$

The pore water pressure at Point A consists of the hydrostatic pore water pressure (u_s) and additional pore water pressure (Δu) caused by seepage,

$$u = u_s + \Delta u$$

where $u_s = 8 \cdot 9.81 = 78.5 \text{kN m}^{-2}$

To find Δu, we will determine the hydraulic gradient (i) for the clay layer (L is the thickness of clay layer, which is equal to 6 m)

$$i = \frac{H}{L} = \frac{5-4}{6} \approx 0.17$$

The difference in heads (Δh_p) at Point A which would cause the upward seepage from Point A to the top of the clay layer ($L_A = 4$ m) is

$$\Delta h_p = i \cdot L_A = 1.7 \cdot 4 = 0.67 \text{ m}$$

The excess pore water pressure (Δu) at Point A equals

$$\Delta u = 0.67 \cdot 9.81 \approx 6.54 \text{kN m}^{-2}$$

Then, the pore water pressure at Point A will be

$$u = 78.5 + 6.54 \approx 85.1 \text{kN m}^{-2}$$

The effective stress at Point A,

$$\sigma' = \sigma - u = 105.9 - 85.1 = 20.8 \text{kN m}^{-2}$$

b) This problem can be solved using the critical hydraulic gradient (i_c):

$$i_c = \frac{\gamma_{sat} - \gamma_w}{\gamma_w} = \frac{16.7 - 9.81}{9.81} \approx 0.7.$$

The hydraulic gradient that already exists in the clay layer is defined as

$$i = \frac{\Delta h}{L} = \frac{1}{6}$$

The critical conditions would occur when

$$i = i_c$$

Then

$$\frac{\Delta h}{6} = 0.7$$

Pumping will decrease the water level in the lake. Let's assume that the value of x (in *m*) is associated with the onset of quick conditions. Then,

$$\frac{1+x}{6} = 0.7$$

Resulting in

$$x \approx 3.2 \, \text{m}$$

Considering that the current level is 4 m, the critical conditions would occur when the water level drops to

$$4 - 3.2 = 0.8 \text{ m}$$

Problem 6.3 Saturated clay is underlain by the saturated sand as shown in Figure 6.10. The sand is under artesian pressure. The clay has bulk density of 1.7 g cm^{-3}. The pore water

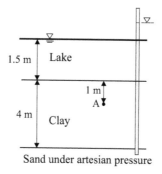

Figure 6.10 Soil profile for Problem 6.3.

pressure at Point A was measured to be 29.4 kN m^{-2}. If water is being drained/pumped out of the lake, the quick conditions may occur in the clay layer.

Estimate the level of water in the lake at which the quick conditions would occur in the clay layer.

Solution

The pore water pressure at Point A consists of the hydrostatic pore water pressure and additional pore water pressure caused by the upward seepage, i.e.,

$$u = u_s + \Delta u$$

The hydrostatic pore water pressure is

$$u_s = (1.5 + 1) \cdot \gamma_w$$

To calculate the additional pore water pressure (Δu), we need to find the hydraulic gradient for the whole layer and changes in heads at Point A as follows:

$$i = \frac{\Delta h}{4} \rightarrow \Delta h = i \cdot l = \frac{\Delta h}{4} \cdot 1$$

Then, the pore water pressure at Point A will be

$$u = 2.5 \cdot \gamma_w + 0.25 \Delta h = 29.4 \, \text{kN m}^{-2}$$

From this relationship, we will get

$\Delta h = 2 \, \text{m}$ (Δh is the difference between the water levels in the lake and in the piezometer).

We will now find the critical hydraulic gradient for the clay layer and use it to determine the water level in the lake at which the critical conditions would occur.

The unit weight of soil, $\gamma = 1.7 \cdot 9.81 \approx 16.68 \text{kN m}^{-3}$

The critical hydraulic gradient, $i_c = \dfrac{\gamma_{sat} - \gamma_w}{\gamma_w} = \dfrac{16.68 - 9.81}{9.81} \approx 0.7$

Assume that x (in m) is the decrease in the water level at which the critical conditions would occur, then

$$i_c = 0.7 = \frac{3.5 - (1.5 - x)}{4} \rightarrow x = 0.8 \, \text{m}$$

6.6 Review quiz

1. The stress acting on the boundary of a soil element is called

   a) total stress
   b) effective stress
   c) pore water pressure

2. Pore water pressure is measured in

   a) N   b) $N\,m^{-2}$ c) $N\,m^{-3}$ d) m

3. A vertical stress of 100 kN m^{-2} is applied to saturated clay and no water drainage is permitted, the excess pore water pressure (in $kN\,m^{-2}$) at t = 0 is

   a) 0   b) very close to 0   c) 50   d) 100

4. The coefficient of permeability has the same units as

   a) hydraulic gradient   b) velocity
   c) rate of seepage   d) it has no units

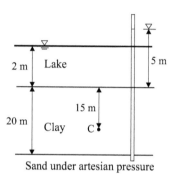

Figure 6.11 Upward seepage through Clay (not to scale).

5. What is the difference in heads (ΔH) that causes the upward seepage (Fig. 6.11)?

   a) 5 m   b) 2 m   c) 2.5 m   d) 3 m

6. Hydraulic gradient at point C (Fig. 6.11) is

   a) 5/15   b) 2/20   c) 3/22   d) 3/20

7. Pressure head is measured in

   a) N   b) $N\,m^{-2}$ c) $N\,m^{-3}$ d) m

8. Hydraulic gradient depends on soil permeability; i.e., the higher the coefficient of permeability, the greater the hydraulic gradient is

   a) True   b) False

9. For most of natural soils, the critical hydraulic gradient is considered to be around

   a) 0   b) 0.25   c) 0.5   d) 1

Answers: 1) a   2) b   3) d   4) b   5) d   6) d   7) d   8) b   9) d

Chapter 7

Principles of water flow in soil

Project relevance: Canal construction involves soil excavation, which would likely cause changes in ground water conditions. This may result in water seepage under the structure, which would have an impact on the stress state in soil mass. This section will introduce the concept of a flow net, which is used to estimate the rate of water flow in soil mass.

7.1 Soil permeability

The rate of water flow (or seepage) is related to hydraulic gradient (i) and soil permeability. The latter is soil property which is described by the *coefficient of permeability (k)*. Note that k has the same units as the velocity (v) of flow (i.e., $m\ s^{-1}$). The coefficient of permeability tends to decrease as fines content increases. Typical ranges of k for different types of soil are summarized in Table 7.1.

Question: *If soil has a very low coefficient of permeability, is it good or bad?*
Answer: In many cases, soils with very low permeability are clays that contain montmorillonite/smectite. Such soils often cause geotechnical problems due to their high plasticity and low strength. As water flows very slowly in such soils, the process of consolidation (removal of water from this soil) would also take much longer than for soils with high permeability such as coarse-grained material. However, there are some positive aspects as well. Soils with very low permeability are used as liners to contain municipal waste as they prevent toxic leachates from contaminating ground water (Gratchev et al., 2014).

Table 7.1 Typical values of the coefficient of permeability (k) for different soils

Type of soil	Permeability (m s^{-1})
Gravel	$1-10^{-1}$
Clean sand	$10^{-2}-10^{-4}$
Fine sand, silt	$10^{-5}-10^{-7}$
Clay	$10^{-7}-10^{-9}$

7.2 Rate of water flow and velocity

For steady water flow through a soil layer with an area of A (Fig. 7.1), we will use the following definitions: a) the rate of flow (q, in m^3 s^{-1}) is defined as the ratio between the quantity of flow (Q, in m^3) and the time (t, in s) necessary to collect this amount (Equation 7.1);

$$q = \frac{Q}{t} \tag{7.1}$$

b) the velocity of flow (in m s^{-1}) is defined as the ratio between the rate of flow (q) and the area (A) of flow (Equation 7.2);

$$v = \frac{q}{A} \tag{7.2}$$

c) the velocity of flow can also be defined using the Darcy's law (Equation 7.3) where it depends on the coefficient of permeability (k) and hydraulic gradient (i).

$$v = k \cdot i \tag{7.3}$$

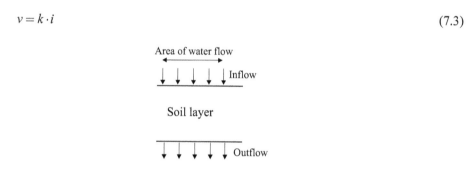

Figure 7.1 Water flow through soil layer with the quantity of flow (Q) over area of A in time t.

7.3 Laboratory tests to determine the coefficient of permeability

The coefficient of permeability is an important parameter used to solve various geotechnical problems related to water. It can be obtained in the laboratory from either constant head or falling head tests. The advantages of these tests are: a) easy to perform as only small amount of soil is required, and b) relatively inexpensive and not time-consuming. However, both tests have shortcomings, as small-sized laboratory samples do not accurately represent the real field conditions where large soil masses are involved. Therefore, care needs to be taken when applying the results from lab tests to characterize site conditions.

Question: *Is there any preference regarding what test to perform?*
Answer: The constant head test is typically used for granular materials (gravel and/or sand) while the falling head test is performed for fine-grained soils.

7.3.1 Laboratory work: constant head test

The constant head test provides a simple way of permeability measurement. The soil sample is contained within a fixed-wall permeameter with inlet and outlets. Water flows one

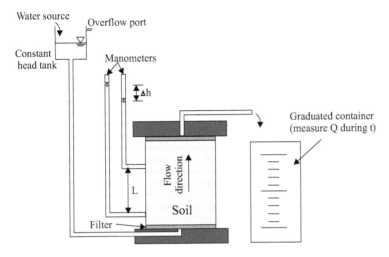

Figure 7.2 Constant-head hydraulic conductivity test configuration.

dimensionally through the sample in the upward direction (Fig. 7.2). The hydraulic gradient is determined from the head difference (Δh) indicated by manometers inserted at two points (L is the distance between the manometers). During testing, the hydraulic gradient (i) is found for different flow rates (q). The flow rate is determined using the amount of water (Q) measured over the time of t. The coefficient of permeability is calculated using the Darcy's law and temperature adjustment is made when necessary.

Question: *Can we test the soil in downward flow?*
Answer: It is possible to test the soil in downward flow; however, it is recommended to use the upward flow as shown in Figure 7.2. In downward flow, the soil tends to become more compacted as the hydraulic conductivity increases.

Test details: Specimen diameter, D = 80 mm   Specimen area, A = 5026 mm^2

Table 7.2 Data from constant head tests

Test No.	*1*	2	3
Manometer port spacing, L (mm)	75	75	75
Time of test, t (sec)	30	30	30
Volume of flow, Q (mm^3)	211.1	482.5	814.2
Discharge, q = Q t^{-1} (mm^3 s^{-1})	7.04	16.1	27.1
Head difference, Δh (mm)	15	30	45
Hydraulic gradient, i = Δh L^{-1}	0.2	0.4	0.6
Coefficient of permeability, k (mm s^{-1})	0.007	0.008	0.009
Average coefficient of permeability, k (mm s^{-1})	0.008		

Note: $k = \dfrac{Q \cdot L}{\Delta h \cdot A \cdot t}$

7.3.2 Laboratory work: falling head test

The falling head test is used to determine the permeability of fine-grained soils. Water flows from a standpipe of cross-sectional area a, through the soil sample which is placed in perme-ameter of cross-sectional area A (Fig. 7.3). At the start of the test (time $t = t_1$), the water level is at a height H_1; the water level then falls as water flows through the soil sample. At the end of the test (time is t_2), the water level is at a height of H_2.

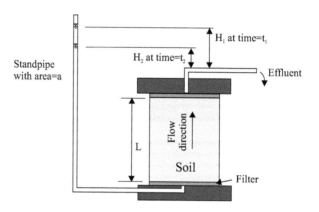

Figure 7.3 Falling-head hydraulic conductivity test configuration.

Table 7.3a Test details.

Specimen diameter: D = 10 cm Specimen area: A = 78.5 cm² Length of soil specimen: L = 12.73 cm
Standpipe diameter: D = 1 cm   Standpipe area: A = 0.785 cm²

Table 7.3b Data from falling head tests

Test No.	1	2	3
Initial reading of standpipe, h_1 (cm)	171.4	171.5	171.7
Final reading of standpipe, h_2 (cm)	156.0	150.6	133.4
Time, t (=$t_1 - t_2$), (s)	1008	1422	2712
Coefficient of permeability, k (cm s⁻¹)	1.18×10^{-5}	1.16×10^{-5}	1.18×10^{-5}
Average coefficient of permeability, k (cm s⁻¹)	1.17×10^{-5}		

Note: $k = \dfrac{a \cdot L}{A \cdot t} \ln\left(\dfrac{h_1}{h_2}\right)$

7.4 Horizontal and vertical water flow in layered soil mass

In practice, we often deal with more complex problems when water flows through several soil layers. In such cases, the following principles of horizontal and vertical water flow are applied (Table 7.4). We will discuss these types of flow in detail while working on Problems 7.2 and 7.3.

Table 7.4 Principles of water flow in layered soil mass

Horizontal flow in layered soil mass	Vertical flow in layered soil mass
The outflow is the same as the inflow	The total head is the sum of the head in each layer
Total flow is the sum of the flow through each layer	The velocity of flow is the same through each layer
Hydraulic gradient is the same through every layer	

The average coefficient of permeability of the layered soil mass depends on the coefficient of permeability of each layer and the layer thickness. It can be calculated for vertical flow (k_v) using Equations 7.4, and horizontal flow (k_h) using Equation 7.5.

$$k_v = \frac{H}{\dfrac{H_1}{k_1} + \dfrac{H_2}{k_2} + \ldots \dfrac{H_n}{k_n}} \tag{7.4}$$

$$k_h = \frac{1}{H}\left[k_1 \cdot H_1 + k_2 \cdot H_2 + \ldots k_n \cdot H_n \right] \tag{7.5}$$

7.5 Elevation, pressure and total heads

Chapter 6.3.2 explained how to calculate additional pore water pressures caused by upward seepage. In this section, we will discuss a different approach to solve similar problems using *total*, *elevation* and *pressure* heads. Water flow in soil between two points is caused by the difference in total heads (Equation 7.6), where the total head (h_t) is defined as the sum of the elevation head (h_e) and the pressure head (h_p).

$$h_t = h_e + h_p \tag{7.6}$$

In geotechnical engineering, the pressure head (h_p) is a very important parameter because it is linked to the pore water pressure (u) as shown in Equation 7.7:

$$u = h_p \cdot \gamma_w \tag{7.7}$$

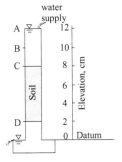

Figure 7.4 Water flow through soil: total, elevation and pressure heads. The head values are given in Table 7.5.

Let's consider soil sample confined in a tube where a steady downward water flow exists (Fig. 7.4). The water only flows through the soil from Point C to Point D due to the difference in total heads between these two points. Let's find the total heads for these two points.

For Point C, the elevation head (h_e) is the distance from Point C to the datum (*reference*) line, which is equal to 8 cm.

Question: *Where do we usually draw the datum line?*
Answer: As the rule of thumb, the datum line is drawn at the lowest water level, which makes the whole analysis much easier.

The pressure head (h_p) indicates how much water is above Point C; that is,

$$h_p = 4\,cm\,(12 - 8 = 4)$$

The total head (h_t) at Point C will be

$$h_t = h_p + h_e = 4 + 8 = 12\,cm$$

Question: *It is clear that the elevation head at Point D will be 2 cm but what will be the pressure head?*
Answer: Remember that water flows only through soil. The total head (h_t) at Point C (where the flow starts) has its maximum value of 12 cm while the total head at Point D (end of water flow) will become zero, as there is no water flow from Point D towards the datum line. Considering that, the pressure head (h_p) at Point D will be

$$h_p = h_t - h_e = 0 - 2 = -2\,cm$$

Table 7.5 summarizes all heads that exist for the experimental setup in Figure 7.4. It is evident from this table that there is no water flow between Points A and C because the total head at these points is the same (12 cm).

Table 7.5 Summary of the heads for the experimental setup in Figure 7.4.

Heads, cm	A	B	C	D
Elevation, h_e	12	10	8	2
Pressure, h_p	0	2	4	-2
Total, h_t	12	12	12	0

7.6 Principles of flow nets

When water flows through soil under engineering structures (for example, a dam), the direction of water flow is not only vertical or horizontal. In such complex cases, the flow net method can be used to estimate the water seepage and pore water pressures. The principles of flow net are briefly described in Figure 7.5.

Flow lines. The water flow path is shown by flow lines. The area confined between two flow lines is called "flow channel". Flow lines are required to be more or less parallel; three

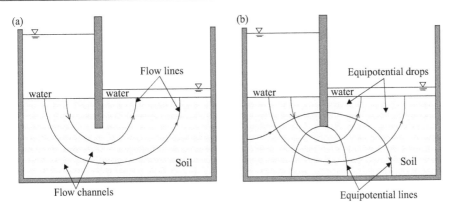

Figure 7.5 Principles of flow net: a) flow lines indicate the direction of water flow; and b) equipotential lines used to connect points with the same total head.

or four flow channels can be sufficient for most of flow nets. For example, there are three flow channels ($N_f = 3$) in the flow net from Figure 7.5a.

Equipotential lines. Lines connecting points with the same total heads are called *equipotential lines*. Equipotential lines must cross flow lines at right angles, forming squares (or rectangles) which are called "equipotential drops". There are five equipotential drops ($N_d = 5$) in each flow channel in Figure 7.5b.

7.6.1 Rate of seepage in flow nets

To calculate the rate of flow/seepage (q in $m^3\ s^{-1}$), Equation 7.8 which relates the total head (H), coefficient of permeability (k), and the geometry of the flow net (N_f – number of flow channels and N_d – number of equipotential drops) is used.

$$q = k \cdot H \cdot \frac{N_f}{N_d} \tag{7.8}$$

Consider a flow net shown in Figure 7.6, where water seepage occurs under a dam in a permeable soil layer. The coefficient of permeability of this soil is $1 \times 10^{-6}\ m\ s^{-1}$. The difference in heads (ΔH) that causes the water seepage under the dam is 18 m, the number of flow

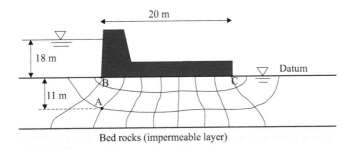

Figure 7.6 Flow net under a dam.

channels (N_f) is 3, and the number of equipotential drops (N_d) is 9. The rate of flow can be estimate using Equation 7.8 as follows:

$$q = k \cdot H \cdot \frac{N_f}{N_d} = 1 \cdot 10^{-6} \cdot 18 \cdot \frac{3}{9} = 6 \cdot 10^{-6} \, \text{m}^3 / \text{s per m run}$$

7.6.2 Pore water pressures in flow nets

There is a procedure to estimate the pore water pressure at any depth using the flow net geometry. The final goal is to obtain the pressure head (h_p) as it is related to pore water pressure (Equation 7.7). The pressure head (h_p) can be found through the difference between the total head (h_t) and elevation (h_e) head. To determine the elevation head, we need to define a datum line (using the lowest level as the datum). For the flow net in Figure 7.6,

$h_e = -11$m (Note that the value of h_e is negative because Point A is below the datum).

As the water flows under the dam from the upstream to the downstream, the total head will gradually dissipate. The greatest value of total head is at the ground level (on the left) before the water flow begins

$$h_t = \Delta H = 18 \text{m}$$

and the lowest value, $h_t = 0$ at the ground level on the right-hand side after passing the last equipotential drop. The total head will decrease by a certain value after passing an equipotential drop. This value is known as the "total head loss" and it is calculated as

$$\Delta h = H / N_d = \frac{18}{9} = 2 \text{m}$$

To reach Point A, the water flow would pass two equipotential drops, which will decrease the total head by a value of

$$2 \cdot \Delta h = 2 \cdot 2 = 4 \text{m}$$

The total head at Point A will be

$$h_t = H - \Delta h \cdot 2 drops = 18 - 4 = 14 \text{m}$$

The pressure head at Point A is estimated as

$$h_p = h_t - h_e = 14 - (-11) = 25 \text{m}$$

Finally, the pore water pressure at Point A equals

$$u = h_p \cdot \gamma_w = 25 \cdot 9.81 \approx 245.3 \text{ kN m}^{-2}$$

7.6.3 Hydraulic uplift force under structures

When there is a steady water flow under large engineering structures like dams, the water seepage will create an uplift force that can undermine the stability of the structure. The uplift force (U, *in kN m⁻¹ per unit length of dam*) depends on the pore water pressure underneath the dam, which is measured along the dam's foundation (Equation 7.9).

$$U = \gamma_w \cdot \left[\frac{h_{p1} + h_{p2}}{2} \cdot L_{1,2} \right]$$ (7.9)

where h_{p1} and h_{p2} are the pressure heads at the edge of the dam, $L_{1,2}$ is the distance between Points 1 and 2.

Let's calculate the uplift force acting on the dam in Figure 7.6. We will first find the pressure heads at Points B and C and then use Equation 7.9 to determine U. The elevation head (h_e) for both points B and C is 0 m. To reach Point B, the water flow would pass 0.7 equipotential drop, which will decrease the total head by a value of

$$2 \cdot \Delta h = 1.4\,\text{m}$$

The total head at Point B is

$$h_t = 18 - 1.4 = 16.6\,\text{m}$$

The pressure head at Point B will equal

$$h_p = h_t - h_e = 16.6 - 0 = 16.6\,\text{m}.$$

For Point C (it takes about 8.2 equipotential drops to reach Point C), the total head is

$$18 - 8.2 \cdot 2 = 1.6\,\text{m}$$

The pressure head will be
$$h_p = 1.6 - 0 = 1.6\,\text{m}$$

The uplift force is estimated as

$$U = 9.81 \cdot \left[\frac{16.6 + 1.6}{2} \cdot 20 \right] = 1785.4\,\text{kN m}^{-1} \text{ per unit length of dam.}$$

7.7 Project analysis: flow net

During canal excavation, sheet piles will be used to provide the stability of the construction site (Fig. 7.7). It is expected that after the excavation, the water seepage would occur in the alluvium layer. There is concern that the underground seepage may significantly increase the pore water pressure in the alluvium, thus undermining the stability of the whole soil mass. We are required to (a) estimate the rate of seepage under the sheet pile, (b) calculate the pore water pressure at Point A and (c) obtain the hydraulic gradient and velocity of flow (in m s⁻¹) at Point A (assume that the width of the equipotential drop for Point A is 2 m). The coefficient of permeability is 6.6×10^{-10} m s⁻¹.

Question: *Was there any test performed to obtain such a low value of the coefficient of permeability?*

Answer: Yes, but this test was conducted using an oedometer (see Section 10.6.3 for details). The oedometer test is carried out to determine the compression characteristics of soil; however, its results can also be used to estimate soil permeability (see Section 10.6.3 for details).

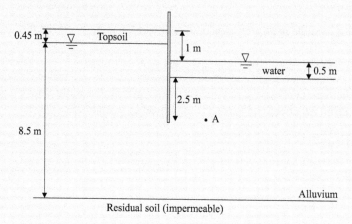

Figure 7.7 Site conditions after the pile sheet installation.

Solution

We will draw a flow net for the site conditions as shown in Figure 7.8, where $N_f = 4$ and $N_d = 8$.
The total head is the difference between two water levels

$$H = 1 - 0.45 = 0.55\,\text{m}$$

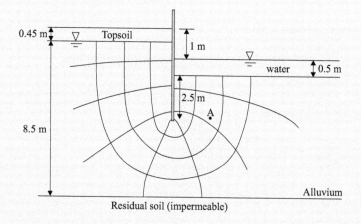

Figure 7.8 Flow net for the project site.

a) The rate of flow, $q = k \cdot H \cdot \dfrac{N_f}{N_d} = 6.6 \cdot 10^{-10} \cdot 0.55 \cdot \dfrac{4}{8} \cdot 86400 \approx 0.000015\,\text{m}^3\ \text{day}^{-1}$

b) The elevation head, $h_{el} = -(0.5 + 2.5) = -3\,\text{m}$

The total head loss for each equipotential drop equals

$$\Delta h = \frac{H}{N_d} = \frac{0.55}{8} = 0.07\,\mathrm{m}$$

As it takes about 6.2 drops to Point A, the total head at Point A will be

$$h_t = 0.55 - 6.2 \cdot 0.07 = 0.12\,\mathrm{m}$$

The pressure head at Point A equals

$$h_p = h_t - h_{el} = 0.12 - (-3) = 3.12\,\mathrm{m}$$

The pore water pressure at Point A will be

$$u = h_p \cdot \gamma_w = 3.12 \cdot 9.81 \approx 30.6\,\mathrm{kN\,m^{-2}}$$

c) The hydraulic conductivity, $i = \dfrac{\Delta h}{L} = \dfrac{0.07}{2} = 0.034$

where L is estimated to be 2 m. The velocity of water flow can be calculated as $v = k \cdot i = 6.6 \cdot 10^{-10} \cdot 0.034 = 2.3 \cdot 10^{-11}\,\mathrm{m\,s^{-1}}$

Question: When I practiced this example by myself, I drew a flow net with five flow channels (N_f = 5) not four, is my answer incorrect?
Answer: Your answer may be slightly different but it doesn't mean it is wrong. Remember that when we draw the flow net, the equipotential drops should be more or less of a square shape. If you use more flow lines, it means that you should also increase the number of equipotential lines to keep the square shape of the drops. As a result, the ratio between N_f and N_d would not be very different from what was shown in Figure 7.8.

7.8 Problems for practice

Problem 7.1 Water flows under constant head through two soil samples as shown in Figure 7.9. The cross-sectional area of the sample is 2200 mm². In 4 minutes, 800 ml water flows through the samples.

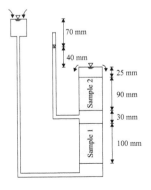

Figure 7.9 Constant water flow through Samples I and 2.

a) Determine the coefficient of permeability (k, in $cm\ s^{-1}$) of Sample 1 and Sample 2
b) If the unit weight of Sample 2 is 20 kN m^{-3}, what will be the effective vertical stress at a point located 30 mm below the top of Sample 2?

Solution

The rate of water flow is

$$q = \frac{Q}{t} = \frac{800}{240} \approx 3.33 cm^3\ s^{-1}$$

The water flow velocity is determined as

$$v = \frac{q}{A} = \frac{3.33}{(2200/100)} \approx 0.152 cm\ s^{-1}$$

The difference in heads (ΔH) for Sample 1 is 70 mm and L = 100 mm. The hydraulic gradient is

$$i = \frac{\Delta H}{L} = \frac{70}{100} = 0.7$$

The coefficient of permeability equals

$$k = \frac{v}{i} = \frac{0.152}{0.7} \approx 0.22 cm\ s^{-1}$$

For Sample 2, we know that ΔH = 40 mm and L = 90 mm, resulting in

The hydraulic gradient, $i = \dfrac{\Delta H}{L} = \dfrac{40}{90} \approx 0.44$

The coefficient of permeability, $k = \dfrac{0.152}{0.44} \approx 0.35 cm\ s^{-1}$

b) The total stress at 30 mm below the top of Sample 2 will be

$$\sigma = \frac{25}{1000} \cdot 9.81 + \frac{30}{1000} \cdot 20 \approx 0.845\ kN\ m^{-2}$$

The total pore water pressure consists of the hydrostatic (u_{static}) and additional pore water pressure (Δu) caused by the upward seepage,

$$u = u_{static} + \Delta u$$

where

$$u_{static} = \frac{55}{1000} \cdot 9.81 = 0.54\ kN\ m^{-2}$$

The additional pore water pressure can be calculated as

$$\Delta u = i \cdot l \cdot \gamma_w = 0.44 \cdot \frac{30}{1000} \cdot 9.81 \approx 0.13 kN\ m^{-2}$$

If the total pore water pressure is

$$u = 0.54 + 0.13 \approx 0.67 \, \text{kN m}^{-2}$$

Then the effective stress will be

$$\sigma' = \sigma - u = 0.845 - 0.67 = 0.175 \, \text{kN m}^{-2}$$

Problem 7.2 A steady water flow is established through a layered soil mass as shown in Figure 7.10. The inflow of water through all the layers was measured to be 90 × 10⁻³ m³ per 10 seconds. For Soil layer 1, the rate of flow through this layer was 20 × 10⁻⁴ m³ s⁻¹, the velocity of the flow was 2 × 10⁻⁴ m s⁻¹ and the coefficient of permeability was 4 × 10⁻⁴ m s⁻¹. For Soil layer 2, the rate of flow was 40 × 10⁻⁴ m³ s⁻¹. Determine:

a) Coefficient of permeability (in $m \, s^{-1}$) for Soil layer 3.
b) Average coefficient of permeability for the layered soil mass.

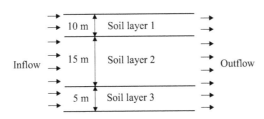

Figure 7.10 Soil profile with steady horizontal water flow.

Solution

Field measurements show that 90 m³ of water passes through the layered soil mass within 10 s. This, the inflow rate (q_{in}) is equal to

$$q_{in} = \frac{Q}{t} = \frac{90 \cdot 10^{-3}}{10} = 9 \cdot 10^{-3} \, \text{m}^3 \, \text{s}^{-1}$$

For horizontal flow in layered soil mass, the inflow is equal to outflow (Table 7.4), i.e.,

$$q_{in} = q_{out} = 9 \cdot 10^{-3} \, \text{m}^3 \, \text{s}^{-1}$$

From the Darcy's law, the hydraulic gradient in Soil layer 1 will be

$$i_1 = \frac{v_i}{k_1} = \frac{2 \cdot 10^{-4}}{4 \cdot 10^{-4}} = 0.5$$

For horizontal flow (Table 7.4) we assume that

$$i_1 = i_2 = i_3 = 0.5$$

Considering that the total flow (Table 7.4) is the sum of the water flow through each layer ($q_{out} = q_1 + q_2 + q_3$), q_3 can be found as

$$q_3 = 90 \cdot 10^{-4} - 20 \cdot 10^{-4} - 40 \cdot 10^{-4} = 30 \cdot 10^{-4} \, \text{m}^3 \, \text{s}^{-1}$$

As we deal with a soil cross-section, the width of each layer is generally taken as 1 m. Then, the area (A_3) of water flow in Soil layer 3 will be

$$A_3 = 5 \cdot 1 = 5 \text{m}^2$$

The velocity of water flow, $v_3 = \dfrac{30 \cdot 10^{-4}}{5} = 6 \cdot 10^{-4} \, \text{m} \, \text{s}^{-1}$

The coefficient of permeability, $k_3 = \dfrac{v_3}{i} = \dfrac{6 \cdot 10^{-4}}{0.5} = 12 \cdot 10^{-4} \, \text{m} \, \text{s}^{-1}$

Similarly, for Soil layer 2,

$$A_2 = 15 \cdot 1 = 15 \text{m}^2$$

The velocity of water flow equals, $v_2 = \dfrac{40 \cdot 10^{-4}}{15} \approx 2.7 \cdot 10^{-4} \, \text{m} \, \text{s}^{-1}$

The coefficient of permeability (k_2), $k_2 = \dfrac{v_2}{i} = \dfrac{2.6 \cdot 10^{-4}}{0.5} = 5.3 \cdot 10^{-4} \, \text{m} \, \text{s}^{-1}$

b) The average coefficient of permeability for horizontal flow in layered soil mass,

$$k_h = \frac{1}{H} \left[k_1 \cdot H_1 + k_2 \cdot H_2 + \ldots k_n \cdot H_n \right]$$

$$= \frac{1}{30} [4 \cdot 10^{-4} \cdot 10 + 5.3 \cdot 10^{-4} \cdot 15 + 12 \cdot 10^{-4} \cdot 5 = 6 \cdot 10^{-4} \, \text{m} \, \text{s}^{-1}$$

Problem 7.3 A sand layer is under artesian pressure as shown in Figure 7.11. The water depth in the lake is 1 m. Assume a steady upward vertical flow through Soil layer 2 and Soil layer 1. For *Soil layer 1*: the coefficient of permeability is 1×10^{-4} m s^{-1} and the hydraulic gradient is 0.2. Determine:

a) Coefficient of permeability (in m s^{-1}) for Soil layer 2
b) Average coefficient of permeability for the layered soil mass

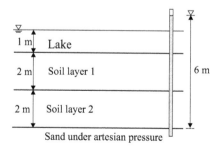

Figure 7.11 Soil profile for Problem 7.3.

Solution

The difference in heads that causes upward seepage is equal to

$$H = 6 - (2+2+1) = 1\,m$$

For a steady vertical flow, the total head is the sum of the head in each layer, i.e.,

$$H = h_1 + h_2$$

For Soil layer 1, we will have

$$i_1 = \frac{h_1}{L_1}$$

Resulting in $h_1 = i_1 \cdot L_1 = 0.2 \cdot 2 = 0.4\,m$

Then, for Soil layer 2, we will get

$$h_2 = 1 - 0.4 = 0.6\,m$$

$$i_2 = \frac{h_2}{L_2} = \frac{0.6}{2} = 0.3$$

Keeping in mind that the velocity of flow is the same for each layer, we will have

$$v = k_1 \cdot i_1 = k_2 \cdot i_2 = 1 \cdot 10^{-4} \cdot 0.2 = 0.2 \cdot 10^{-4}\,m\,s^{-1}$$

The coefficient of permeability for Soil layer 2 can be found as

$$k_2 = \frac{v}{i_2} = \frac{0.2 \cdot 10^{-4}}{0.3} = 0.67 \cdot 10^{-4}\,m\,s^{-1}$$

b) The average coefficient of permeability for vertical flow in layered soil mass,

$$k_v = \frac{H}{\dfrac{H_1}{k_1} + \dfrac{H_2}{k_2} + \ldots \dfrac{H_n}{k_n}} = \frac{4}{\dfrac{2}{1 \cdot 10^{-4}} + \dfrac{2}{0.67 \cdot 10^{-4}}} = 0.8 \cdot 10^{-4}\,m\,s^{-1}$$

Problem 7.4 A sheet pile is driven into sandy silt. The coefficient of permeability of the soil is 1.6×10^{-6} m s^{-1}. Using the flow net shown in Figure 7.12, compute the following:

a) Flow rate in $m^3\,day^{-1}$ per meter run
b) Pore water pressure at Point A
c) Pore water pressure at Point B
d) Hydraulic gradient and velocity of flow (in $m\,s^{-1}$) at point B

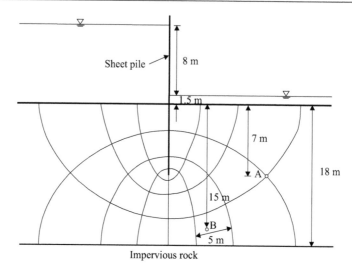

Figure 7.12 Flow net for Problem 7.4.

Solution

For the flow net in Figure 7.12, we know

$$N_f = 4, N_d = 8, H = 8\,\text{m}, k = 1.6 \cdot 10^{-6}\,\text{m s}^{-1}$$

a)   The rate of flow, $q = k \cdot H \cdot \dfrac{N_f}{N_d} = 1.6 \cdot 10^{-6} \cdot 8 \cdot \dfrac{4}{8} \cdot 86400 \approx 0.55\,\text{m}^3\,\text{day}^{-1}$ per m run

b)   The elevation head, $h_e = -8.5\,\text{m}$

The total head loss for each equipotential drop equals $\Delta h = \dfrac{8}{8} = 1\,\text{m}$

As it takes 7 drops to reach Point A, the total head at Point A is calculated as

$$h_t = H - 7 \cdot \Delta h = 1\,\text{m}$$

From Equation 7.6, we get

$$h_p = h_t - h_e = 1 - (-8.5) = 9.5\,\text{m}$$

Finally, the pore water pressure at Point A is

$$u = h_p \cdot \gamma_{water} = 9.5 \cdot 9.81 \approx 93.2\,\text{kN m}^{-2}$$

c)   The same procedure is used to calculate the pore water pressure at Point B

The elevation head (h_e) is -16.5 m,

The total head loss for each equipotential drop equals

$$\Delta h = \dfrac{8}{8} = 1\,\text{m}$$

It takes about 5.4 drops to reach Point B, then the total head at Point B is

$$h_t = H - 5.4 \cdot \Delta h = 2.6\,\text{m}$$

The pressure head at Point B is equal to

$$h_p = h_t - h_e = 2.6 - (-16.5) = 19.1\,\text{m}$$

The pore water pressure at Point B will be

$$u = h_p \cdot \gamma_w = 19.1 \cdot 9.81 \approx 187.4\,\text{kN m}^{-2}$$

d) The hydraulic gradient at Point B

$$i = \frac{\Delta h}{L} = \frac{1}{5} = 0.2$$

where Δh is the total head loss for each equipotential drop, and L is the length of the equipotential drop where Point B is located.
The velocity of water flow at Point B,

$$v = k \cdot i = 1.6 \cdot 10^{-6} \cdot 0.2 = 0.32 \cdot 10^{-6}\ \text{m s}^{-1}$$

Problem 7.5 A dam is shown in Figure 7.13. The coefficient of permeability (k) of the soil is 1.0×10^{-6} m s^{-1}.

a) Sketch a flow net and compute the flow rate (q) in m^3/day per meter run
b) Find the pore water pressure at Point A
c) Calculate the uplift force under the dam

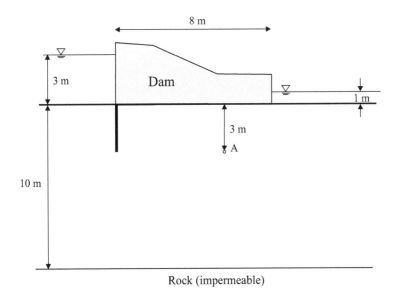

Figure 7.13 Schematic illustration of the dam and site conditions.

Solution

Flow net is shown in Figure 7.14.

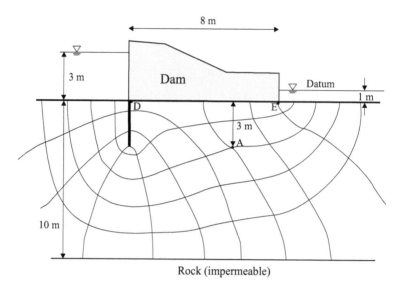

Figure 7.14 Flow net for the site conditions shown in Figure 7.13.

a) The flow net has the following characteristics:

$$N_f = 5, N_d = 10, H = 3-1 = 2\,\mathrm{m}$$

The rate of flow can be calculated as:

$$q = k \cdot H \cdot \frac{N_f}{N_d} = 1 \cdot 10^{-6} \cdot 2 \cdot \frac{5}{10} \cdot 86400 \approx 0.086\,\mathrm{m^3/day\ per\ m\ run}$$

b) To calculate the pore water pressure at Point A, we should find the pressure head (h_p) using Equation 7.6. The elevation head (h_e) and the total head loss (Δh) are:

$$h_e = -4\,\mathrm{m}, \ \Delta h = \frac{2}{10} = 0.2\,\mathrm{m}$$

As there are 7 drops to Point A, the total head at Point A will be equal to

$$h_t = 2 - 7 \cdot 0.2 = 0.6\,\mathrm{m}$$

From Equation 7.6, the pressure head at Point A is

$$h_p = h_t - h_e = 0.6 - 4 = 4.6\,\mathrm{m}$$

The pore water pressure at Point A (equation 7.7) equals

$$u = 4.6 \cdot 9.81 \approx 45.1\,\mathrm{kN\ m^{-2}}$$

c)   To calculate the uplift force (U) under the dam, we should first find the pressure heads at Points D and E and then use Equation 7.9. For Point D,

The elevation head is $h_e = -1\,\text{m}$

It takes about 6.2 drops to reach Point D, thus the total head at Point D will be

$h_t = 2 - 6.2 \cdot 0.2 = 0.76\,\text{m}$

The pressure head equals

$h_p = 0.76 - (-1) = 1.76\,\text{m}$

For Point E,

The elevation head is $h_e = -1\,\text{m}$

9 drops to Point E, resulting in the total head at this point to be

$h_t = 2 - 9 \cdot 0.2 = 0.2\,\text{m}$

The pressure head is

$h_p = 0.2 - (-1) = 1.2\,\text{m}$

The length (L) of the dam is 8 m, the uplift force acting under the dam equals

$$U = \gamma_w \cdot \left[\frac{h_{pD} + h_{pE}}{2} \cdot L_{DE} \right] = 9.81 \cdot \left[\frac{1.76 + 1.2}{2} \cdot 8 \right] \approx 116.2\,\text{kN per m}$$

7.9 Review quiz

1.   Water flows through soil because of the difference in

a)   pressure heads   b)   elevation heads   c)   total heads

2.   Which of the following statements is NOT correct?

a)   Flow lines should be parallel, when possible;
b)   Flow lines typically should create three or four channels;
c)   Flow lines must cross the equipotential lines at right angles;
d)   Flow lines are lines with the same total head.

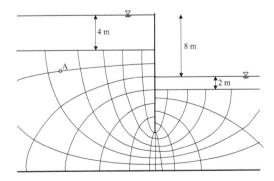

Figure 7.15 Flow net.

3. How many flow channels are in the flow net (Fig. 7.15)?

   a) 5   b) 6   c) 7   d) 12

4. How many equipotential drops are in this flow net?

   a) 10   b) 12   c) 15   d)   24

5. What is the difference (in *m*) in total head that causes the water flow under the pile sheet?

   a) 4   b) 2   c) 11   d) 8

6. What is the total head loss for each equipotential drop?

   a) 8/7   b) 7/12   c) 4/7   d) 8/12

7. What is the total head (in *m*) at Point A?

   a) 7.57   b) 7.62   c) 7.51   d) 7.33

Answers: 1) c   2) d   3) c   4) b   5) d   6) d   7) d

Chapter 8

Mohr circle and stresses

Project relevance: The proposed sand embankment will generate additional loads on the soft and wet alluvial clay. It is thus essential to know how the applied loads will distribute in the soil mass and whether the soil will have sufficient strength to withstand it. This chapter will introduce the Mohr circle concept which is used to estimate stresses acting on a soil element.

8.1 Theoretical considerations

Normal and shear stresses applied to a soil element will determine the amount of deformation that this soil may undergo due to construction loading. Figure 8.1 shows a soil element under a general state of stress. Normal stresses (σ_x and σ_y) act perpendicular to the soil face while shear stresses (τ) act parallel to the face. To distinguish between the stresses, a system of double subscripts (for example, τ_{xy}) is used: the first subscript denotes the direction of the normal to the plane on which the stress acts while the second subscript denotes the direction in which the stress acts.

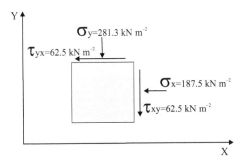

Figure 8.1 Soil element under a state of stress.

Question: *In real life, a soil element will be under a three-dimensional (3D) state of stress (σ_x, σ_y and σ_z). Why do we only consider a (2D) cross-section of the soil element?*
Answer: In many geotechnical problems, there is no need to conduct a three-dimensional stress analysis because many structures including embankments and retaining walls are long in comparisons with their width and height. Three-dimensional analysis is also more complex, yet it can be performed by existing geotechnical software.

There are a few important points that need to be considered:

- In soil mechanics, normal (compressive) stresses and anticlockwise shear stresses are positive. For example, the normal stress (σ_y) acting on the top of the soil element in Figure 8.1 is positive 281.25 kN m^{-2} while the shear stress (τ_{xy}) is negative (–)62.5 kN m^{-2}.

Question: *Why is it different from structural mechanics where tensile normal stresses and clockwise shear stresses are conventionally taken as positive?*
Answer: The reason for this is that soils cannot sustain tensile stresses; that is why in soil mechanics, we adopt that the "compression is positive".

- Under a state of stress, there are always two perpendicular planes at which the shear stress (τ) is equal to zero. Such planes are called the "principal planes" while the normal stresses acting on these planes are called *principal stresses*.
- The largest principal stress is termed the *major* principal stress (σ_1) while the other one is called the *minor* principal stress (σ_3). For field conditions, it is commonly assumed that the normal vertical stress (σ_v) is the major principal stress ($\sigma_1 = \sigma_v$) while the normal horizontal stress (σ_h) is the minor principal stress ($\sigma_3 = \sigma_h$).

8.2 Mohr circle of stress

When normal and shear stresses are known, the Mohr circle can be drawn as shown in Figure 8.2 for the stress conditions given in Figure 8.1.

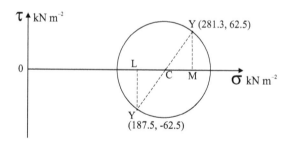

Figure 8.2 Mohr circle for the soil element from Figure 8.1. Explanation is given in the text.

The drawing procedure used in Figure 8.2 involves few steps (a-d) as detailed below:

a) $LC = CM = ((OM - OL))/2 = ((281.3 - 187.5))/2 = 46.88\text{kN m}^{-2}$

b) $R = \sqrt{XL^2 + LC^2} = \sqrt{YM^2 + CM^2} = \sqrt{62.5^2 + 46.88^2} = 78.13\text{kN m}^{-2}$

c) $OC = 234.4\text{kN m}^{-2}$

d) $\sigma_1 = OC + R = 234.4 + 78.13 = 312.5\text{kN m}^{-2}$

$\sigma_3 = OA = OC - R = 234.4 - 78.13 = 156.3\text{kN m}^{-2}$

From the aforementioned procedure, we can derive Equations 8.1 and 8.2, which are used to find the principle stresses without having to draw the Mohr circle.

$$R = \sqrt{\left(\frac{\sigma_x - \sigma_y}{2}\right)^2 + \tau_{xy}^2} \tag{8.1}$$

$$\sigma_{1,3} = \left(\frac{\sigma_x + \sigma_y}{2}\right) \pm R \tag{8.2}$$

8.3 Determining stresses acting on plane

To determine the normal (σ_n) and shear (τ_n) stresses acting on any plane, Equations 8.3 and 8.4 are used.

$$\sigma_n = \frac{\sigma_y + \sigma_x}{2} + \frac{\sigma_y - \sigma_x}{2}\cos 2\theta + \tau_{xy}\sin 2\theta \tag{8.3}$$

$$\tau_n = \frac{\sigma_y - \sigma_x}{2}\sin 2\theta - \tau_{xy}\cos 2\theta \tag{8.4}$$

Let's see how we can use these equations to find the stresses acting on a plane inclined at 45° in Figure 8.3.

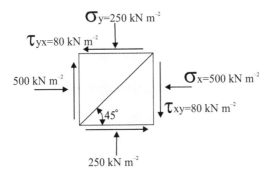

Figure 8.3 Normal and shear stresses acting on a soil element.

From Figure 8.3, we have

$\sigma_y = 250$ kN m^{-2}, $\sigma_x = 500$ kN m^{-2}, $\tau_{xy} = -80$ kN m^{-2}, $\theta = 45°$

Using Equations 8.3 and 8.4, we get

$$\sigma_n = \frac{250 + 500}{2} + \frac{250 - 500}{2}\cos\left(2 \cdot 45°\right) - 80 \cdot \sin\left(2 \cdot 45°\right) \approx 295\text{kN m}^{-2}$$

$$\tau_n = \frac{250 - 500}{2}\sin\left(2 \cdot 45°\right) - (-80) \cdot \cos\left(2 \cdot 45°\right) \approx -125\text{kN m}^{-2}$$

Question: *How do we determine stresses in practice? Do we use this method to manu-ally calculate stresses acting at different depths? It seems to be very laborious and time-consuming job.*
Answer: In practice, engineers use geotechnical software which will greatly minimize the time involved. However, it is still important to understand the theory behind it.

8.4 Pole method

The Pole method is a graphical solution to obtain the normal and shear stresses acting on any plane. The letter "P" is typically used to identify Pole on the Mohr circle. Once Pole is identified, a line parallel to the plane of interest can be drawn from Pole that will intersect the Mohr circle at a point that gives the normal and shear stresses acting on the plane of inter-est. A procedure to identify Pole for the stress conditions shown in Figure 8.4 as well as the normal and shear stresses acting on EF is discussed below.

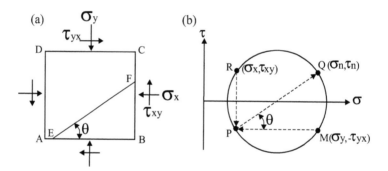

Figure 8.4 The Pole method. Point P denotes Pole.

Procedure

1. Locate Points R (σ_x, τ_{xy}) and M (σ_y, $-\tau_{yx}$) in the "σ vs. τ" space (Fig. 8.4b).
2. Draw the Mohr circle.
3. To locate Pole, draw a line MP through Point M (σ_y, $-\tau_{yx}$), which is parallel to the plane on which the stresses of Point M act. It will be a horizontal line from M to P (Fig. 8.4b) because σ_y and τ_{yx} act on the horizontal plane DC (Fig. 8.4a).
4. Intersection of MP with the circle is Pole (Point P).
5. Draw a line PQ from P, which is parallel to EF.
6. Determine the coordinates of Point Q, which are the normal and shear stresses acting on EF.

Question: *Do we always use the combination of σ_y and τ_{yx} to locate Pole?*
Answer: It is not necessary, we can also use σ_x and τ_{xy} instead. For example, in Figure 8.4b, we can draw a vertical line parallel to CB from Point R (σ_x, τ_{xy}) and identify the point where this line crosses the circle. It will be the same point P (Pole).

8.5 Project analysis: Mohr circle and stresses in soil mass

Let's analyze the stresses acting on Point C located in the soft clay 2 m below the ground after the embankment construction (Fig. 8.5a). Assume that the additional load from the embankment would produce stresses as shown in Figure 8.5b. We will calculate the normal (σ_n) and shear (τ_n) stresses acting on a plane AB using the mathematical equations and Pole method.

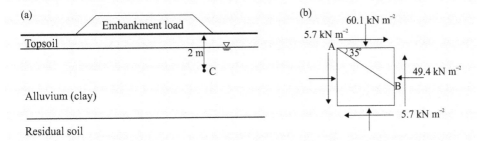

Figure 8.5 Location of Point C in the alluvium (a); stresses acting on Point C (b).

Solution

From Figure 8.5b, we know

$$\sigma_y = 60.1 \text{ kN m}^{-2}, \sigma_x = 49.4 \text{ kN m}^{-2}, \tau_{xy} = 5.7 \text{ kN m}^{-2}, \theta = 145°$$

Using Equations 8.3 and 8.4, we obtain

$$\sigma_n = \frac{60.1+49.4}{2} + \frac{60.1-49.4}{2}\cos\left(2\cdot145°\right) + 5.7\cdot\sin\left(2\cdot145°\right) \approx 51.2\text{kN m}^{-2}$$

$$\tau_n = \frac{60.1-49.4}{2}\sin\left(2\cdot145°\right) - 5.7\cdot\cos\left(2\cdot145°\right) \approx -6.9\text{kN m}^{-2}$$

The graphical solution using the Pole method is given in Figure 8.6.

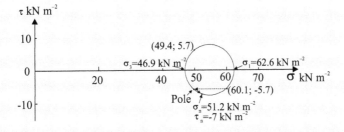

Figure 8.6 Using the Pole method to find the stresses acting on Plane AB from Figure 8.5b. The solution is: the normal (σ_n) stress is 51.2 kN m^{-2} and the shear (τ_n) stress is −7 kN m^{-2}.

8.6 Problems for practice

Problem 8.1 For the soil element in Figure 8.7,

a) Draw the Mohr circle and locate Pole
b) Determine the values of principal stresses (σ_1 and σ_3)
c) Determine the normal and shear stresses on plane AB

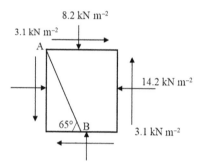

Figure 8.7 Stresses acting on a soil element in Problem 8.1.

Solution

a) The graphical solution using the Pole method is shown in Figure 8.8

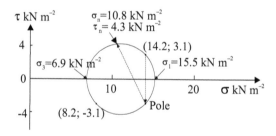

Figure 8.8 Mohr circle for the stress condition in Problem 8.1.

b) To find the principal stresses, we will calculate the radius (R) of the Mohr circle using Equation 8.1,

$$R = \sqrt{\left(\frac{\sigma_x - \sigma_y}{2}\right)^2 + \tau_{xy}^2} = \sqrt{\left(\frac{14.2 - 8.2}{2}\right)^2 + 3.1^2} = 4.3\text{kN m}^{-2}$$

Using Equation 8.2, the principal stresses will be as follows:

$$\sigma_{1,3} = \left(\frac{\sigma_x + \sigma_y}{2}\right) \pm R \rightarrow \sigma_1 \approx 15.5\text{kN m}^{-2}; \sigma_3 \approx 6.9\text{kN m}^{-2}$$

c)   From Figure 8.7, we know

$$\sigma_x = 14.2\text{kN m}^{-2}, \sigma_y = 8.2\text{kN m}^{-2}, \tau_{xy} = 3.1\text{kN m}^{-2}, \theta = 115°$$

Using Equation 8.3, we will find the normal stress acting on Plane AB,

$$\sigma_n = \frac{\sigma_y + \sigma_x}{2} + \frac{\sigma_y - \sigma_x}{2}\cos 2\theta + \tau_{xy}\sin 2\theta$$

$$= \frac{8.2+14.2}{2} + \frac{8.2-14.2}{2}\cos\left(2\cdot 115°\right) + 3.1\cdot\sin\left(2\cdot 115°\right) \approx 10.75\text{kN m}^{-2}$$

Using Equation 8.4, we will find the shear stress acting on Plane AB,

$$\tau_n = \frac{\sigma_y - \sigma_x}{2}\sin 2\theta - \tau_{xy}\cos 2\theta = \frac{8.2-14.2}{2}\sin\left(2\cdot 115°\right) - 3.1\cdot\cos\left(2\cdot 115°\right) \approx 4.29\text{kN m}^{-2}$$

Problem 8.2 For the soil element shown in Figure 8.9:

a)   Draw the Mohr circle and locate Pole
b)   Determine the normal and shear stresses on plane AB

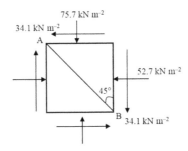

Figure 8.9 Stresses acting on a soil element in Problem 8.2.

Solution

a)   The graphical solution including the Pole method is shown in Figure 8.10.

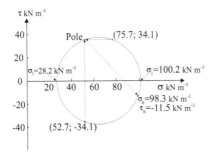

Figure 8.10 Mohr circle for the stress condition in Problem 8.2.

b) From Figure 8.9, we get

$$\sigma_x = 52.7\text{kN m}^{-2}, \sigma_y = 75.7\text{kN m}^{-2}, \tau_{xy} = -34.1\text{kN m}^{-2}, \theta = 135°$$

The normal stress acting on Plane AB,

$$\sigma_n = \frac{\sigma_y + \sigma_x}{2} + \frac{\sigma_y - \sigma_x}{2}\cos 2\theta + \tau_{xy}\sin 2\theta$$

$$= \frac{75.7 + 52.7}{2} + \frac{75.7 - 52.7}{2}\cos\left(2 \cdot 135°\right) - 34.1 \cdot \sin\left(2 \cdot 135°\right) \approx 98.3\text{kN m}^{-2}$$

The shear stress acting on Plane AB,

$$\tau_n = \frac{\sigma_y - \sigma_x}{2}\sin 2\theta - \tau_{xy}\cos 2\theta = \frac{75.7 - 52.7}{2}\sin\left(2 \cdot 135°\right) - \left(-34.1 \cdot \cos\left(2 \cdot 135°\right)\right) \approx -11.5\text{kN m}^{-2}$$

Chapter 9

Principles of soil deformation

Project relevance: Once we know the stresses acting in soil mass, we can estimate the amount of deformation that soil can undergo during construction after the load application. This chapter will introduce the definitions of strain and discuss laboratory techniques used to measure soil strength and deformation.

9.1 Soil deformation in practice

Unlike rocks, soil is a relatively weak material and higher stresses acting in soil mass can cause significant deformations. In practice, this frequently results in differential settlements of engineering structures and their subsequent collapse. Figure 9.1 shows the failure of sand embankment built on very soft clay in Thailand.

Figure 9.1 Collapse of the soft Bangkok clay during excavation for a flood drain canal near the Suvarnabhumi International Airport (Courtesy of Dr Noppadol).

9.2 Laboratory tests to study soil strength

To estimate soil deformation characteristics, we conduct laboratory tests including uncon-fined or triaxial compression where we load soil specimens until failure. These tests provide us with the knowledge of soil strength, a parameter that determines the capacity of soil to deform under stresses. Oedometer tests (see Chapter 10) can directly determine the magni-tude of deformation that occurs in soil under different loads.

9.2.1 Unconfined compression test

This test is a fast and easy-to-perform experiment that yields the strength characteristics of clayey soil. The axial stress (σ_1) is applied to the specimen until failure (Fig. 9.2). During this test, the soil strength and deformation are recorded. Unfortunately, the experimental setup does not represent the stress conditions that exist in the field (such as vertical and horizontal stresses as shown in Figure 6.1b) because it is not possible to apply the confining pressure ($\sigma_3 = 0$, Figure 9.2a) in this test.

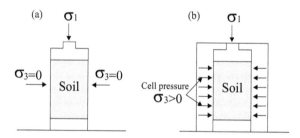

Figure 9.2 Stresses acting on soil specimen in: a) unconfined compression test, and b) triaxial compres-sion test.

Question: *Does this test work for all soils?*
Answer: No, this test is performed only for clayey soils as granular materials (sand or gravel) are not capable of keeping cylindrical shape during testing. Sand samples are usually tested in triaxial compression (Fig. 9.2b).

9.2.2 Triaxial compression test

Triaxial compression tests (Fig. 9.3) enable more accurate simulation of field stress condi-tions in the laboratory by considering the confining pressure ($\sigma_3 > 0$). However, it is a more expensive and time-consuming experiment compared to the unconfined compression. This test gives the strength characteristics of soil which are used in design. In addition, triaxial tests can be conducted on both fine-grained and coarse-grained materials, and can repro-duce drained and undrained conditions. More details about this test and analysis of obtained results will be provided in Chapter 11.

9.2.3 Drained versus undrained tests

Drained compression tests are performed in a way that no excess pore water pressure can be generated during loading because the water in soil can freely drain out. Such

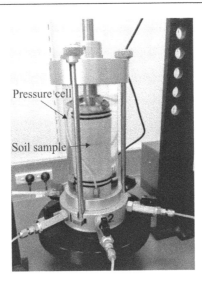

Figure 9.3 A triaxial compression test: cell pressure and soil sample.

tests are conducted to simulate the behavior of granular material in the field under loads when proper drainage is put in place. It is also performed to estimate the strength of soil over a long period of time when the excess pore water pressure is expected to fully dissipate. In contrast, when fine-grained soils (especially saturated clay) are loaded, the excess pore water pressure would rapidly generate, creating undrained conditions. As water flows rather slowly in fine-grained soils due to their extremely low permeability, the excess pore water pressure would dissipate very slowly, and thus the undrained conditions would remain in soil for a long time. The undrained conditions should be avoided in engineering practice because the shear strength of soil under undrained conditions is significantly lower compared to the soil strength under drained conditions (See Section 11 for details).

9.3 Stress-strain characteristics of soil

9.3.1 Definition of strain

Soil deformations are measured in strain. Axial (ε_1) and radial (ε_3) strains are defined as shown in Figure 9.4 and Equations 9.1 and 9.2.

$$\Delta\varepsilon_1 = \frac{l_0 - l}{l_0} \tag{9.1}$$

$$\Delta\varepsilon_3 = \frac{r_0 - r}{r_0} \tag{9.2}$$

where l_0 *and l* – the length of sample before and after the load application, respectively; r_0 *and r* – the radius of sample before and after the load application, respectively.

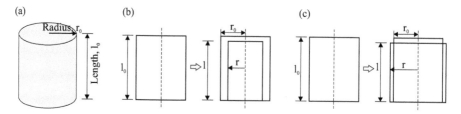

Figure 9.4 Definition of soil deformation (strain): a) cylindrical soil sample before load application; b) drained deformations of soil sample; c) undrained deformations of soil sample.

In drained compression tests on saturated soil, water drainage is permitted, and as a result, the volume of soil and water content generally decrease. In contrast, during undrained compression of saturated soil, i.e., when water drainage is *not* permitted, the volume of soil sample remains constant (Fig. 9.4c).

Using the Poisson's ratio (v) and Young's modulus (E), it is possible to calculate the strain ($\Delta\varepsilon_1$ and $\Delta\varepsilon_3$) in soil caused by changes in the principal stresses ($\Delta\sigma_1$ and $\Delta\sigma_3$). For drained conditions (Equations 9.3 and 9.4), we will use the drained Young's modulus (E') and Poisson's ratio (v'):

$$\Delta\varepsilon_1 = \frac{1}{E'}\left[\Delta\sigma_1' - v'\left(2\Delta\sigma_3'\right)\right] \tag{9.3}$$

$$\Delta\varepsilon_3 = \frac{1}{E'}\left[\Delta\sigma_3' - v'\left(\Delta\sigma_1' + \Delta\sigma_3'\right)\right] \tag{9.4}$$

For undrained conditions, we will use the undrained Young's modulus (E_u) and Poisson's ratio (v_u) as shown in Equations 9.5 and 9.6:

$$\Delta\varepsilon_1 = \frac{1}{E_u}\left[\Delta\sigma_1' - v_u\left(2\Delta\sigma_3'\right)\right] \tag{9.5}$$

$$\Delta\varepsilon_3 = \frac{1}{E_u}\left[\Delta\sigma_3' - v_u\left(\Delta\sigma_1' + \Delta\sigma_3'\right)\right] \tag{9.6}$$

Question: *Is the Poisson's ratio (v) the same for drained and undrained conditions?*
Answer: No, it is different. For saturated clay, the undrained Poisson's ratio (v_u) is commonly taken as 0.5. For drained conditions, v' typically varies from 0.2 to 0.3.

Under loads, soil can also experience volumetric (ε_v) and shear (ε_s) deformations that are estimated using Equations 9.7 and 9.8:

$$\varepsilon_v = \varepsilon_1 + 2\varepsilon_3 \tag{9.7}$$

$$\varepsilon_s = \frac{2}{3}(\varepsilon_1 - \varepsilon_3) \tag{9.8}$$

9.3.2 Pore water pressure coefficients

During loading of saturated soil mass, especially clay, the increase in the principal stresses ($\Delta\sigma_1$ and/or $\Delta\sigma_3$) results in the generation of excess pore water pressure (Δu). The relationship between the principal stresses and excess pore water pressure is given in Equation 9.9, where two coefficients (A and B) are used to describe the properties of soil and degree of saturation, respectively.

$$\Delta u = B\left[\Delta\sigma_3 + A\left(\Delta\sigma_1 - \Delta\sigma_3\right)\right] \tag{9.9}$$

Coefficient B depends on the degree of saturation, and it reaches its maximum value (B=1) when soil is fully saturated. Coefficient A depends on the type of soil, and its value typically varies from 0.4 to 1.

9.4 Project analysis: deformation of soft clay after load application

Let's calculate changes in the stresses and estimate deformation of the soft alluvial clay after the application of load. Let's assume that the sand embankment would generate the additional normal stress (q) of about 63.3 kN m^{-2} as shown in Figure 9.5. We will find the stresses at Point D before and after the load application, and use this information to estimate the relevant soil deformation. The following data are provided for the alluvial clay layer: $K_0 = 0.9$, $E_u = 1000$ kN m^{-2} and A = 0.4.

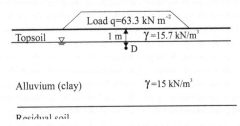

Figure 9.5 Estimation of soil strain at Point D.

Solution

Before the load application, the total stress at Point D equals

$$\sigma_v = 0.45 \cdot 15.7 + 0.55 \cdot 15 = 15.3 \text{kN m}^{-2}$$

The pore water pressure,

$$u = 0.55 \cdot 9.81 = 5.4 \text{kN m}^{-2}$$

The effective vertical stress,

$$\sigma'_v = 15.3 - 5.4 = 9.9 \text{kN m}^{-2}$$

The effective horizontal stress,

$$\sigma'_h = \sigma'_v \cdot K_0 = 9.9 \cdot 0.9 \approx 8.9 \text{kN m}^{-2}$$

The total horizontal stress,

$$\sigma_h = 8.9 + 5.4 \approx 14.3 \text{kN m}^{-2}$$

After the load application, the vertical stresses ($\Delta\sigma_v$) will increase by q (63.3 kN m^{-2}) and the horizontal stress will increase ($\Delta\sigma_h$) by 0.9q (56.9 kN m^{-2}). The pore water pressure will increase by value of Δu, which is estimated using Equation 9.9, where $\Delta\sigma_1 = \Delta\sigma_v$, and $\Delta\sigma_3 = \Delta\sigma_h$.

$$\Delta u = B\left[\Delta\sigma_3 + A\left(\Delta\sigma_1 - \Delta\sigma_3\right)\right] = 1[56.9 + 0.4(63.3 - 56.7) = 59.5 \text{kN m}^{-2}$$

Question: *Why do we assume that B=1?*
Answer: It is reasonable to assume that the undrained conditions would exist in satu-rated soft clay immediately after the load application since there is no sufficient time for this low permeability clayey soil to rapidly dissipate the excess pore water pressures.

Note that the increase in pore water pressure (Δu) is almost as much as the load (q) applied. This implies that most of the applied load is being carried by the water in soil.
To estimate soil deformation at Point D, we will refer to Equation 9.5 to calculate the axial strain:

$$\Delta\varepsilon_1 = \frac{1}{E_u}\left[\Delta\sigma_1 - \nu_u \cdot 2 \cdot \sigma_3\right] = \frac{1}{1000}\left[63.3 - 0.5 \cdot 2 \cdot 56.9\right] = 0.006$$

9.5 Problems for practice

Problem 9.1 In a drained triaxial test on a cylindrical specimen of sand with an initial height of 75 mm and diameter 37.5 mm, the reduction in the height was found to be 3.0 mm and the reduction in the volume was 1.645 ml.
Calculate the axial strain, volumetric strain, radial strain and shear strain of the specimen.

Solution

Figure 9.6 gives the experimental setup used in the drained triaxial test.

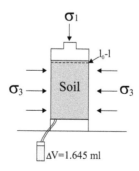

Figure 9.6 Schematic illustration of the experimental setup in Problem 9.1

The initial sample volume, $V = \pi \dfrac{D^2}{4} H = \dfrac{3.14}{4} \cdot \left(\dfrac{37.5}{1000} \right)^2 \cdot \dfrac{75}{1000} = 82,85 \cdot 10^{-6} \, \text{m}^3$

Using the definition of volumetric strain, we will find

$$\Delta\varepsilon_v = \dfrac{\Delta V}{V} = \dfrac{1.645 \cdot 10^{-6}}{82.85 \cdot 10^{-6}} = 0.02$$

From the definition of axial strain, we will obtain

$$\Delta\varepsilon_1 = \dfrac{l - l_1}{l} = \dfrac{3}{75} = 0.04$$

By re-arranging Equation 9.7, we get

$$\Delta\varepsilon_r = \dfrac{1}{2}\left(\Delta\varepsilon_v - \Delta\varepsilon_1 \right) \approx -0.01$$

The shear strain is calculated using Equation 9.8 as

$$\Delta\varepsilon_s = \dfrac{2}{3}\left(\Delta\varepsilon_1 - \Delta\varepsilon_3 \right) \approx 0.033$$

Problem 9.2 A foundation generates a uniform load (q) of 150 kN m^{-2}. A layer of clay follows a 3-m thick sand layer, with the ground water table located 1.5 m below the ground surface (Fig. 9.7).

Sand: The unit weight of sand above the ground water table is 18 kN m^{-3} and below the ground water table is 21 kN m^{-3}.

Clay: The unit weight of clay is 20 kN m^{-3}. The coefficient of earth pressure at rest for the clay is 0.9. The Skempton's A parameter is 0.4. The drained Young's modulus is 5000 kN m^{-2} and the undrained Young's modulus is 2000 kN m^{-2}.

According to the elasticity theory, the stresses at Point B change as follows:

$$\Delta\sigma_v = 0.8 \cdot q, \text{ and } \Delta\sigma_h = 0.23 \cdot q$$

a) Calculate the total and effective vertical and horizontal stresses at Point B immediately after the loading (assume undrained conditions)
b) Calculate the vertical strain at Point B due to the *undrained* loading

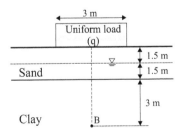

Figure 9.7 A cross-section of the soil mass from Problem 9.2.

Solution

Before loading, the total vertical stress at Point B equals

$$\sigma_v = 1.5 \cdot 18 + 1.5 \cdot 21 + 3 \cdot 20 = 118.5 \text{kN m}^{-2}$$

The pore water pressure, $u = 4.5 \cdot 9.81 = 44.15 \text{kN m}^{-2}$

The effective vertical stress, $\sigma'_v = 118.5 - 44.15 = 74.35 \text{kN m}^{-2}$

The effective horizontal stress, $\sigma'_h = \sigma'_v \cdot K_0 = 74.35 \cdot 0.9 \approx 66.92 \text{kN m}^{-2}$

The total horizontal stress, $\sigma_h = 66.92 + 44.15 \approx 111.1 \text{kN m}^{-2}$

After the load application, the stresses will increase as follows:
The vertical stress will increase by

$$\Delta\sigma_v = 0.8 \cdot q = 0.8 \cdot 150 = 120 \text{kN m}^{-2}$$

The horizontal stress will increase by

$$\Delta\sigma_h = 0.23 \cdot q = 34.5 \text{kN m}^{-2}$$

The pore water pressure will increase, according to Equation 9.9,

$$\Delta u = B\left[\Delta\sigma_3 + A\left(\Delta\sigma_1 - \Delta\sigma_3\right)\right] = 1[34.5 + 0.4(120 - 34.5) = 68.7 \text{kN m}^{-2}$$

New total stresses immediately after the load application are:

$$\sigma_v = 118.5 + 120 = 238.5 \text{kN m}^{-2}$$

$$\sigma_h = 111.07 + 34.5 = 145.57 \text{kN m}^{-2}$$

$$u = 44.15 + 68.7 = 112.85 \text{kN m}^{-2}$$

New effective stresses immediately after the load application are:

$$\sigma'_v = 238.5 - 112.85 = 125.66 \text{kN m}^{-2}$$

$$\sigma'_h = 145.57 - 112.85 = 32.72 \text{kN m}^{-2}$$

b) Using Equation 9.5, we will get

$$\Delta\varepsilon_1 = \frac{1}{E_u}\left[\Delta\sigma'_1 - \nu_u \cdot 2 \cdot \sigma'_3\right] = \frac{1}{2000}[120 - 0.5 \cdot 2 \cdot 34.5] = 0.043$$

9.6 Review quiz

1. Shear stress that causes clockwise shear is positive

    a) True   b) False

2. In Figure 9.8, Pole is located in Point:

    a) B   b) Z   c) L   d) X

3. For saturated clay, it is commonly assumed that

    a) A≈1   b) B≈1   c) A≈B≈1   d) no assumptions can be made

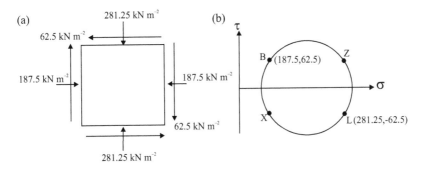

Figure 9.8 Stresses acting on a soil element (a) and the Mohr circle (b).

4. For undrained conditions, the Poisson's ratio of saturated clay is commonly assumed to be

    a) 0.2   b) 0.25   c) 0.3   d) 0.5

5. The type of deformations at which the volume of saturated soil sample remains constant occurs under

    a) drained conditions
    b) undrained conditions
    c) both drained and undrained conditions
    d) never occurs in soil

6. What statement is not correct?

    a) For saturated clay, the Skempton's B parameter is commonly assumed 1
    b) The Poisson's ratio of saturated clay under undrained conditions is commonly assumed to be 0.5
    c) In drained triaxial compression tests, the moisture content of saturated clay sample will typically decrease when the load is applied
    d) In undrained triaxial compression tests, the volume of saturated clay sample will typically decrease when the load is applied

7. Figure 9.9a shows the principal stresses acting on the soil element. The normal and shear stresses acting on the plane AB are represented by the coordinates of Point (Fig. 9.9b):

    a) C   b) D   c) E   d) F

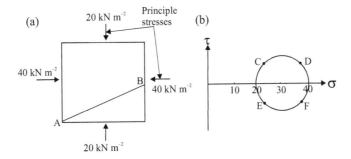

Figure 9.9 Principle stresses acting on a soil element (a) and the Mohr circle (b).

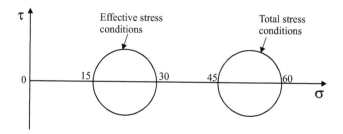

Figure 9.10 Test results given for effective and total stress conditions. All values are given in kN m⁻².

8. Test results are analyzed in terms of total and effective stress conditions (Fig. 9.10). What is the value of pore water pressure (kN m⁻²)?

a) 15   b) 30   c) 45   d) 60

Answers: 1) b   2) a   3) b   4) d   5) b   6) d   7) c   8) b

Chapter 10

Consolidation of soft soil

Project relevance: Previous analysis indicated that immediate construction on the very soft alluvium would be unwise as this soil layer is very loose and saturated. The properties of this soil need to be improved to make it more appropriate foundation. This chapter will introduce the principles of soil consolidation and discuss techniques commonly used in engineering practice to calculate the amount of soil settlements and the time necessary to accomplish this task.

10.1 Process of consolidation

When saturated clay is loaded, most of the applied load ($\Delta\sigma$) will be transferred to the water in soil, thus generating excess pore water pressures ($\Delta u \approx \Delta\sigma$). Over time (which can take days, months or even years), the excess pore water pressure will eventually dissipate ($\Delta u \rightarrow 0$) as the pressurized pore water pressure permeates into an adjacent layer of soil with higher permeability. As this occurs, $\Delta\sigma$ will be gradually transferred from the pore water pressure to the soil particles, and the soil mass will undergo some settlements. When the excess pore water pressure has completely dissipated, the effective stress will increase by $\Delta\sigma$ ($\Delta\sigma' \approx \Delta\sigma$). This time-dependent process is known as consolidation.

Question: *How is consolidation different from soil compaction? In the end of each process, the soil will become denser with a higher strength anyway?*
Answer: Compaction is meant to remove mostly air from coarse-grained soils while consolidation decreases the water content in fine-grained soils (mostly clays).

Figure 10.1 demonstrates what typically occurs in the field when soft and saturated clay (about 6 m thick) is consolidated by means of a pre-load method. To initiate the process of consolidation, an embankment was constructed that generated additional stresses (about 55 kN m^{-2} as shown in Figure 10.1a). The pore water pressure was measured at some depth in the soft clay layer during and after the construction stage. It indicated rapid increases in the excess pore water pressure (about 48 kN m^{-2}, Figure 10.1b) immediately after the embankment construction was completed. As the drainage was permitted, the excess pore water pressure began to dissipate, gradually decreasing to its initial value of 4 kN m^{-2}. This time-consuming process continued for more than 1000 days.

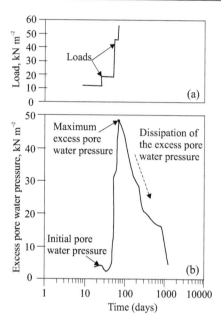

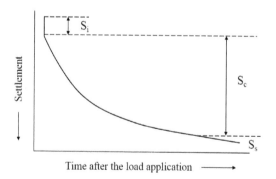

Figure 10.1 Field data on soft soil consolidation caused by the embankment load (a) and changes in the pore water pressure with time (b). The site conditions and process of consolidation are described in detail in (Gratchev *et al.*, 2012).

10.2 Types of settlements during consolidation

In engineering practice, a pre-load method is commonly used to consolidate soft clay. When the load is applied, three types of settlements can occur (Fig. 10.2):

Figure 10.2 Different types of settlements after the application of load: S_i – immediate deformation, S_c – consolidation and S_s – secondary settlements.

1) *Immediate* settlements (S_i) when air and some water at the surface escape from the soil immediately after the load application. The amount of settlement during this stage is typically very small. 2) *Consolidation* (or primarily) (S_c) settlements are much larger in

magnitude and time-consuming. The water is gradually forced out of the soil as the applied stress is slowly transferred to soil particles, resulting in soil settlements. 3) *Secondary* settlements (S_s) are related to viscous or creep deformation of soil structure. It may take a long time to develop and it is associated with limited settlement of soil.

Question: *What is the engineering approach to deal with consolidation in practice?*
Answer: There are two important questions which engineers are required to find solutions for: "What will be the maximum settlement?" and "How long will it take?" This can be done by testing soil samples in the laboratory by means of oedometer. The following sections will outline the testing procedure and discuss how to calculate soil settlements and time of consolidation.

10.3 Soil consolidation in practice

Unfortunately, the pre-load method can take a significant period of time (years) as the water drains out of soil very slowly due to the low permeability of clay. Such delay is often not acceptable for construction projects. To speed up the process of consolidation, PVDs (pre-fabricated vertical drains made of geofabric) can be used. A PVD is porous inside which allows water to flow from the clay mass to the PVD, and then through the PVD to the ground surface. PVDs are installed in soil mass prior to embankment construction to shorten the water flow path (compare the water path in Figure 10.3a and Figure 10.3b), thus significantly decreasing the time of consolidation. Note that although PVDs accelerate the process of consolidation, the final settlement will still be the same as the one where no drains are employed.

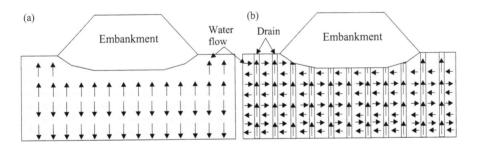

Figure 10.3 A pre-load method without PVDs (a) and with PVDs (b).

Question: *Are there any other methods to accelerate the consolidation process of soil?*
Answer: When engineers first faced this problem, wick drains were used to consolidate soft clays. Several boreholes were drilled in soft clay deposits and then filled with granular material such as sand before a pre-load was applied. It provided good results; however, it proved to be very costly. PVDs have proven to be as effective and, more importantly, significantly cheaper (Indraratna et al., 1994).

Figure 10.4 A pre-load method with PVDs in Port of Brisbane (Australia). The embankment was used to consolidate very soft marine deposits. PVDs with a spacing of about 1.5 m were employed to provide drainage for water.

Question: *From Figure 10.3, it seems that the spacing between PVDs will affect the time of consolidation. If we place PVDs closer to each other (let's say 0.5 m), will it accelerate consolidation even more?*

Answer: Practice and research (Bergado et al., 1991; Indraratna and Redana, 1998) show that it is not a good idea due to the "smear" zone effect. When PVDs are inserted in the ground, the soil around them becomes disturbed, which resulted in even lower coefficients of permeability. This will slow down the water flow in the soils, making the process of consolidation more time-consuming.

10.4 One-dimensional consolidation

Consolidation of soft clay occurs as the void space of soil decreases. For this reason, the settlement of clay layer (ΔH) in the field can be related to changes in the soil void ratio (Δe) as shown in Equation 10.1.

$$\frac{\Delta H}{H_0} = \frac{\Delta e}{1+e_0} \qquad (10.1)$$

where ΔH is the consolidation settlement, H_0 is the initial thickness of clay layer, Δe is the change in void ratio of soil and e_0 is the initial void ratio. This relationship indicates that the settlement of clay in the field can be estimated on the basis of laboratory data where only small amount of soil is tested. This has a great advantage, as it will minimize the cost of geotechnical investigation.

Question: *How reliable are the results of laboratory tests considering that only a small soil sample is tested? Will it accurately represent the field conditions?*

Answer: There is always concern with applying results from laboratory tests to solve large-scaled field problems. Engineering practice shows that the results obtained in the laboratory may be very different from the field data. The problem arises from the size of soil samples tested; i.e., small laboratory samples often do not account for the presence of larger voids or inclusions of other material, which is common in the field (Gratchev et al., 2012).

10.4.1 Laboratory consolidation tests

To estimate the settlement of clay in the field, a series of consolidation tests are performed in the laboratory by means of oedometer (Fig. 10.5). During this test, clay a sample (generally, 6 cm in diameter and 2 cm in height) is loaded using prescribed stress increments. For each stress increment, changes in the sample height are monitored and recorded within at least 24 h. The obtained results plotted and analyzed to obtain the compression index (C_c) and the coefficient of volume compressibility (m_v).

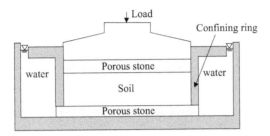

Figure 10.5 A schematic oedometer test setup. The soil specimen is confined in the ring while load is applied vertically. The soil is kept in water to prevent drying and shrinkage during testing.

Laboratory data from oedometer tests are plotted as the effective stress against void ratio using a log scale (Fig. 10.6). The compression index (C_c) is defined for a large range of stresses, and it is obtained from the linear and steep part of the experimental curve.

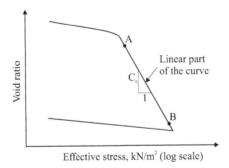

Figure 10.6 Definition of the compression index (C_c) using laboratory data from an oedometer test.

Considering Point A (σ'_1, e_1) and Point B (σ'_2, e_2) in Figure 10.6, the compression index C_c is calculated as follows (Equation 10.2):

$$C_c = \frac{(e_1 - e_2)}{log\sigma'_2 - log\sigma'_1}$$
(10.2)

In contrast, the coefficient of volume compressibility (m_v) is defined for a much narrow range of stresses using an arithmetic scale for the applied stress (Fig. 10.7).

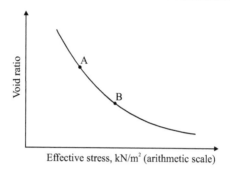

Figure 10.7 Definition of the coefficient of volume compressibility (m_v) using laboratory data from an oedometer test.

This makes m_v rather sensitive to changes in stresses, and its value can significantly vary depending on the stress range. Considering two points: Point A (σ'_0, e_0) and Point B (σ'_1, e_1) on the curve in Figure 10.7, the coefficient of volume compressibility is defined as

$$m_v = \frac{\Delta e / (1 + e_1)}{\sigma'_1 - \sigma'_2}$$

(10.3)

10.4.2 Calculations of total settlements

To estimate the total settlement (S_c) of soft clay, the compression index (C_c) (Equation 10.4) or the coefficient of volume compressibility (m_v) (Equation 10.5) can be used.

$$S_c = \frac{C_c \cdot H}{1 + e_0} log \frac{\sigma'_0 + \Delta \sigma}{\sigma'_0}$$

(10.4)

where H is the initial thickness of the clay layer, σ_0' is the initial effective stress in the middle of the clay layer, $\Delta \sigma$ is the change in the effective stress due to additional load.

$$S_c = m_v \cdot \Delta \sigma' \cdot H$$

(10.5)

where H is the initial thickness of the clay layer, $\Delta \sigma'$ is the change in the effective vertical stress that causes consolidation. The following section will discuss in detail how to use these equations.

10.5 Project analysis: soil consolidation

We will now analyse the total settlement in the soft alluvial clay which would occur due to the placement of 3-m sand fill (embankment) as schematically shown in Figure 10.8. The sand will be taken from Pit 1, and it will be compacted to its maximum dry density and optimum water content. We will use the following information: the density of topsoil is 1.60 g cm⁻³, and the average density of the alluvial clay is 1.53 g cm⁻³.

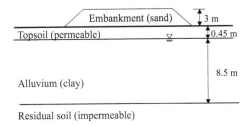

Figure 10.8 Schematic illustration of a pre-load method at the project site.

Let's estimate the total settlement of the alluvium clay after the embankment construction and the time necessary to complete 90% consolidation.

Question: *Is there any reason for selecting 90% consolidation?*
Answer: It is commonly assumed that the process of consolidation is over when the degree of consolidation (U) reaches 90%, and thus, in many cases, this is when the construction stage can begin.

Solution

We will determine the compression index of the alluvial clay and calculate the total settlement using Equation 10.4. Soil samples were collected from a depth of 2 m to study the compression characteristics of the soft clay. The results from oedometer tests are given in Tables 10.1 and 10.2.

10.5.1 Laboratory work: oedometer test

The experimental data from Table 10.2 are plotted in Figure 10.9 with a semi-log scale to obtain the compression index of the soil.
    By considering two points (A and B) on the linear part of the experimental curve (Fig. 10.9), the compression index (C_c) can be calculated as follows:

$$C_c = \frac{1.96 - 1.1}{log\,384 - log\,48} \approx 0.95$$

Table 10.1 Test details

Material: High plasticity alluvial clay	Initial moisture content: 80.9%
Initial dry density: 1.51 t m^{-3}	Sample height: 17 mm
Soil particle density: 2.53 t m^{-3}	Sample diameter: 45 mm

Table 10.2 Data from oedometer test

Normal stress, kN m^{-2}	12	24	48	96	192	384	96	24
Void ratio	2.06	2.03	1.96	1.68	1.37	1.10	1.15	1.18

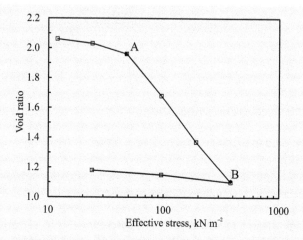

Figure 10.9 Results of oedometer test on very soft clay (Alluvium).

10.5.2 Total settlements after consolidation

When the soil in embankment is compacted to its maximum dry density (ρ_{dmax} = 1.85 g cm^{-3}) and optimum water content (w_{opt} = 16.8%), its bulk density will be:

$$\rho_{dry} = \frac{\rho}{1+w} \to 1.85 = \frac{\rho}{1+0.168} \to \rho \approx 2.16\text{g cm}^{-3}$$

The unit weight will be $\gamma \approx 21.2\text{kN m}^{-3}$, and the embankment will generate an additional effective stress ($\Delta\sigma'$), which equals

$$\Delta\sigma' = 21.2 \cdot 3 \approx 63.6\text{kN m}^{-2}$$

The unit weight of each soil is shown in Figure 10.10.

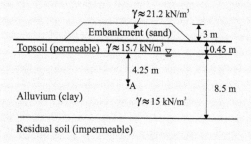

Figure 10.10 Soil properties used in the analysis.

To calculate the total settlement, we will use Equation 10.4 and data from the oedometer test (Fig. 10.9) to a) estimate the initial effective stresses acting in the middle of the clay layer (Point A in Figure 10.10) prior to the embankment construction, and b) determine the void ratio (e_0) that corresponds to this effective stress.

The effective stress (σ_v') at Point A prior to the embankment construction is:

$$\sigma = 0.45 \cdot 15.7 + 4.25 \cdot 15 \approx 70.9 \text{kN m}^{-2}$$

$$u = 4.25 \cdot 9.81 \approx 41.7 \text{kN m}^{-2}$$

$$\sigma_v' = 70.9 - 41.7 = 29.2 \text{kN m}^{-2}$$

It is evident from Figure 10.9 that that the void ratio of about 2.0 would correspond to the effective stress of 29.2 kN m^{-2}. Using Equation 10.4, we will calculate the total settlement of the clay layer:

$$S_c = \frac{C_c \cdot H}{1 + e_0} \log \frac{\Delta \sigma_0' + \Delta \sigma}{\Delta \sigma_0'} = \frac{0.95 \cdot 8.5}{1 + 2} \log \frac{29.2 + 63.6}{29.2} \approx 1.354 \text{m}$$

10.6 Terzaghi's theory of consolidation and its practical application

Karl Terzaghi was a geotechnical engineer and geologist who is regarded as the "father of soil mechanics". One of his major contributions is the one-dimensional consolidation equation (Eq. 10.6) which describes how the excess pore water pressure (Δu) dissipates with time (t) with depth (z):

$$\frac{\partial u}{\partial t} = c_v \frac{\partial^2 u}{\partial z^2} \tag{10.6}$$

where c_v is the coefficient of consolidation, which is defined in Equation 10.7

$$c_v = \frac{k}{m_v \cdot \gamma_w} \tag{10.7}$$

where k is the coefficient of permeability and m_v is the coefficient of volume compressibility. Terzaghi's theory of consolidation yields Equation 10.8, which is used to estimate the time of consolidation.

$$T_v = \frac{c_v \cdot t}{H_{dr}^2} \tag{10.8}$$

where t is the time to reach a certain degree of consolidation, T_v is the time factor, which is related to the degree of consolidation (U), and H_{dr} is the drainage path of water.

The relationship between U and T_v is given in Table 10.3; for example, for a degree of consolidation (U) of 50%, the time factor (T_{50}) is 0.197.

It is obvious that the process of consolidation can only occur when water drainage is permitted. The availability of drainage on the top and/or bottom of the clay layer is a key factor that affects the time of consolidation. The type of drainage is defined by the parameter H_{dr} as shown in Figure 10.11.

If there is a clay layer of thickness H, which is sandwiched between two permeable layers such as sand or gravel, then $H_{dr} = H/2$ (double drainage). If the drainage is available only at the top or bottom (single drainage) then $H_{dr} = H$.

Table 10.3 Correlations between T_v and U.

U, %	T_v	U, %	T_v	U, %	T_v	U, %	T_v
0	0	26	0.0531	52	0.212	78	0.529
1	0.00008	27	0.0572	53	0.221	79	0.547
2	0.0003	28	0.0615	54	0.230	80	0.567
3	0.00071	29	0.0660	55	0.239	81	0.588
4	0.00126	30	0.0707	56	0.248	82	0.610
5	0.00196	31	0.0754	57	0.257	83	0.633
6	0.00283	32	0.0803	58	0.267	84	0.658
7	0.00385	33	0.0855	59	0.276	85	0.684
8	0.00502	34	0.0907	60	0.286	86	0.712
9	0.00636	35	0.0962	61	0.297	87	0.742
10	0.00785	36	0.102	62	0.307	88	0.774
11	0.0095	37	0.107	63	0.318	89	0.809
12	0.0113	38	0.113	64	0.329	90	0.848
13	0.0133	39	0.119	65	0.340	91	0.891
14	0.0154	40	0.126	66	0.352	92	0.938
15	0.0177	41	0.132	67	0.364	93	0.993
16	0.0201	42	0.138	68	0.377	94	1.055
17	0.0227	43	0.145	69	0.390	95	1.129
18	0.0254	44	0.152	70	0.403	96	1.219
19	0.0283	45	0.159	71	0.417	97	1.336
20	0.0314	46	0.166	72	0.431	98	1.500
21	0.0346	47	0.173	73	0.446	99	1.781
22	0.0380	48	0.181	74	0.461	100	∞
23	0.0415	49	0.188	75	0.477		
24	0.0452	50	0.197	76	0.493		
25	0.0491	51	0.204	77	0.511		

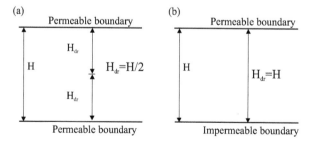

Figure 10.11 Definition of the parameter H_{dr}: a) drainage on both sides (double drainage) and b) drainage only on one side (single drainage).

10.6.1 Coefficient of consolidation

The coefficient of (vertical) consolidation (c_v) is an important parameter used to estimate the time of consolidation. It is obtained from oedometer tests when changes in the sample height caused by the application of load are recorded over time. The obtained data can be analyzed using either the log time method (Fig. 10.12) or root time method (Fig. 10.13).

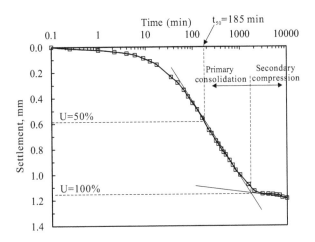

Figure 10.12 Time-settlement data plotted using the log time method. Test conditions: the initial sample height (H) is 100 mm; double drainage (H_{dr} = H/2); the effective stress applied is 25 kN m^{-2}, T_{50} = 0.197.

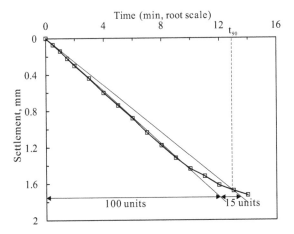

Figure 10.13 Time-settlement data plotted using the root time method. Test conditions: the initial sample height (H) is 20 mm; double drainage (H_{dr} = H/2); the applied effective stress is 12.5 kN m^{-2}, T_{90} = 0.848.

Log time method. The procedure for deriving c_v using the log time method (Fig. 10.12) is as follows: 1) identify the point corresponding to the settlement at 100% consolidation (U = 100%). It is the intersection of the straight portion of the primarily consolidation and secondary compression curves. 2) Identify the point corresponding to the settlement of 50% consolidation, which is in the middle between U = 0% and U = 100%. 3) Identify the time (t_{50}) corresponding to U = 50%. 4) Calculate the coefficient of vertical consolidation (c_v) using Equation 10.8.

For the experimental curve shown in Figure 10.12, the time corresponding to 50% consolidation (t_{50}) is approximately 185 min, which will result in

$$c_v = \frac{T_{50} \cdot H_{dr}^2}{t_{50}} = \frac{0.197 \cdot 0.05^2}{185 \cdot 60} = 4.4 \cdot 10^{-8} \text{ m}^2 \text{ s}^{-1}$$

Root time method. The procedure to derive c_v using the root time method (Fig. 10.13) can be described as follows: 1) Extend the linear portion of the experimental curve to the bottom of the graph. 2) Draw a line between the Y-axis (settlement) and the extension line and assume that its length is 100 units. 3) Along this line, add additional 15 units to the right and identify this point. 4) Connect this point with the graph origin (0) and identify the point of intersection between this new line and the experimental curve. 5) The point of intersection corresponds with t_{90}. 6) Calculate the coefficient of vertical consolidation (c_v) using Equation 10.8.

For the experimental curve shown in Figure 10.13, the time corresponding to 90% consolidation ($\sqrt{t_{90}} = 12.8 \text{ min}$) is approximately 163.8 min, which will give

$$c_v = \frac{T_{90} \cdot H_{dr}^2}{t_{90}} = \frac{0.848 \cdot 0.01^2}{163.8 \cdot 60} = 8.6 \cdot 10^{-9} \text{ m}^2 \text{ s}^{-1}$$

Question: What method should we use?
Answer: Either one is fine; they both give very similar results.

10.6.2 Project analysis: time of consolidation

This analysis suggests that the clay layer would undergo significant deformation (1.35 m) after the application of the embankment load. To find the time necessary to consolidate the clay, we will analyse the results from oedometer test (Table 10.4) using the log time method.

Test data. Sample height: 20 mm Applied stress: 96 kN m^{-2}

Table 10.4 Data from a consolidation test on the soft alluvial clay

Time (min)	0.25	1	4	9	16	25	36	81	1440
Total ΔH (mm)	0.231	0.786	1.824	2.93	3.838	4.02	4.08	4.15	4.28

The time-settlement data from Table 10.4 is plotted in Figure 10.14 using the lot time method. From this graph, we obtain:

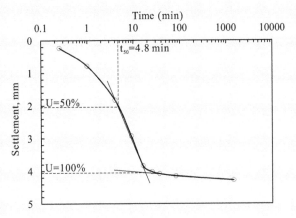

Figure 10.14 Time-settlement data plotted using the log time method.

$t_{50} \approx 4.8\,\mathrm{min}$

As this time is related to U = 50%, then $T_{50} = 0.197$.
In this oedometer test, the drainage is provided on both sides, i.e.

$$H_{dr} = \frac{20}{2} = 10\,\mathrm{mm}$$

The coefficient of consolidation, $c_v = \dfrac{T_{50} \cdot H_{dr}^2}{t_{50}} = \dfrac{0.197 \cdot 0.01^2}{4.8 \cdot 60} = 6.8 \cdot 10^{-8}\,\mathrm{m^2\ s^{-1}}$

To estimate the time to achieve 90% consolidation, we should find the time factor that corresponds to U = 90% (Table 10.3). Therefore, for $U = 90\% \rightarrow T_{90} = 0.848$.

As the residual clay is expected to have rather low permeability, the drainage will be provided only on the top of alluvium layer through the topsoil, meaning that

$$H_{dr} = 8.5\mathrm{m}$$

The time to achieve 90% consolidation in the clay layer of alluvium will be

$$t_{90} = \frac{T_{90} \cdot H_{dr}^2}{c_v} = \frac{0.848 \cdot 8.5^2}{6.8 \cdot 10^{-8}} \approx 895694619\,\mathrm{s} \approx 340.6\,months\,or\,28.4\ years$$

Conclusion: Clearly, 28.4 years is not acceptable for this project, and thus some other methods related to soil consolidation should be considered.

10.6.3 Project analysis: coefficient of permeability

Results from the oedometer test (Fig. 10.9) can be used to obtain the coefficient of permeability of the alluvial clay using Equation 10.7, which can be re-arranged as follows:

$$k = c_v \cdot m_v \cdot \gamma_w$$

As the effective stress in the middle of the clay layer is estimated to be 29.2 kN m^{-2}, we will select two points on the experimental curve (Fig. 10.9), which are close to the value of effective stress. Note that unlike the compression index (C_c), m_v is obtained for a narrow range of normal stresses. In this case, we will use a range of 20 to 40 kN m^{-2}.

From Figure 10.9, we will obtain

$$\sigma_1' = 20\,kN / m^2 \rightarrow e_1 \approx 2.03$$

$$\sigma_2' = 40\,kN / m^2 \rightarrow e_2 \approx 1.97$$

The coefficient of volume compressibility will be

$$m_v = \frac{\Delta e / (1 + e_0)}{\sigma_1' - \sigma_2'} = \frac{(2.03 - 1.97)/(1 + 2.03)}{40 - 20} \approx 0.00099\,m^2\,kN^{-1}$$

The coefficient of permeability equals

$$k = c_v \cdot m_v \cdot \gamma_w = 6.8 \cdot 10^{-8} \cdot 9.9 \cdot 10^{-4} \cdot 9.81 \approx 6.64 \cdot 10^{-10}\,m\,s^{-1}$$

Question: *Does it mean that we can use data from the oedometer test instead of the falling head test to obtain the coefficient of permeability of clay?*
Answer: Yes, according to the theory, these two methods are expected to produce similar results.

10.7 Overconsolidation ratio

Normally consolidated clay, which is very common in engineering practice, means that the clay was consolidated under loads that are currently present in the field. Overconsolidated clay can be found in nature as well. Such clay was previously consolidated under much greater stresses (compared to the current stress). For this reason, overconsolidated clay tends to have a *greater strength* than normally consolidated clays, and in some cases, it may be considered as good engineering material.

The degree of consolidation is determined by the overconsolidation ratio (OCR), which is defined in Equation 10.9.

$$OCR = \frac{\sigma_c'}{\sigma'} \tag{10.9}$$

where σ_c' is the preconsolidation stress and σ' is the present stress.

Normally consolidated clays have OCR=1, while for overconsolidated clays, OCR can be in the range of 2–3.

10.8 Problems for practice

Problem 10.1 The soil profile at a site consists of a 2 m-thick sand layer underlain by a 5 m-thick clay layer, which is underlain by gravel as shown in Figure 10.15. The ground water level is 1 m below the surface.

Sand layer: the unit weight of sand above the ground water level is 16 kN m^{-3}, the unit weight of sand below the ground water level is 18 kN m^{-3}.

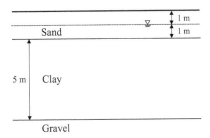

Figure 10.15 Soil profile for Problem 10.1.

Clay layer: A sample was collected from the middle of the Clay layer for laboratory testing. It had bulk density of 1.62 t m^{-2}. The data from an oedometer test on this clay are given in Table 10.5.

Table 10.5 Results from an oedometer test on the clay sample.

Stress, kN m^{-2}	Void ratio, e
6	2.01
12.5	1.98
25	1.94
50	1.76
100	1.58
200	1.35
400	1.125
12.5	1.35

a) Determine the compression index of this clay
b) A 3-m compacted fill with unit weight of 19 kN m^{-3} will be placed at the ground level. What would be the final consolidation settlement (in m) in the clay layer caused by the placement of the fill?
c) A specimen of the clay (a height of 20 mm) was subjected to loading under the stresses that would represent the stress conditions in the clay layer after the placement of the 3-m compacted fill. Data obtained from this consolidation test are given in Table 10.6 in the form of time vs. changes in the specimen height (Total ΔH).

Table 10.6 Data from a consolidation test on the Clay.

Time (min)	0.25	1	4	9	16	25	36	81	1440
Total ΔH (mm)	0.231	0.786	1.824	2.93	3.838	4.02	4.08	4.15	4.28

c1) Using the root time method, determine the coefficient of consolidation (in m s^{-1}) for this clay.
c2) How long (in *months*) will it take for the clay layer to reach 70% consolidation?
c3) For the stress range of 50–100 kN m^{-2}, determine the coefficient of permeability (in *m s^{-1}*) of the clay.

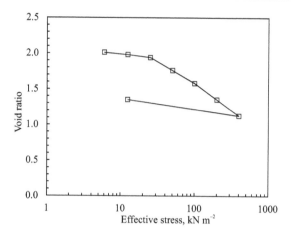

Figure 10.16 Results from an oedometer test on the clay samples (data from Table 10.5).

a)   Results of the oedometer test are plotted in Figure 10.16 as the effective stress against void ratio. From the linear part of the curve, the compression index can be obtained as follows:

$$C_c = \frac{(1.94 - 1.125)}{log\,400 - log\,25} \approx 0.68$$

b)   A cross-section of the soil mass is given in Figure 10.17.

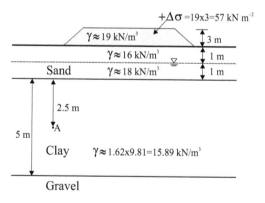

Figure 10.17 Soil profile and soil properties.

To calculate the total settlement of the clay layer, we should find the stresses acting in the middle of this layer (Point A) before the load application:

Normal stress, $\sigma = 1 \cdot 16 + 1 \cdot 18 + 2.5 \cdot 15.9 = 73.7 \text{kN m}^{-2}$

Pore water pressure, $u = (1 + 2.5) \cdot 9.81 = 34.3 \text{kN m}^{-2}$

Effective stress, $\sigma' = \sigma - u = 73.7 - 34.3 = 39.4 \text{kN m}^{-2}$

The initial void ratio (e_0) corresponding to the effective stress of 39.4 kN m^{-2} is approximately 1.8 (from Figure 10.16). Due to the embankment load, the effective stress would increase by $\Delta\sigma' = 57.0$ kN m^{-2} (see Figure 10.17 for explanation).

Final settlement caused by the embankment load, $S_c = \dfrac{0.68 \cdot 5}{1+1.8} log \dfrac{39.4+57}{39.4} \approx 0.47\text{m}$

c1) Results from the oedometer test are analyzed using the root time method (Fig. 10.18).

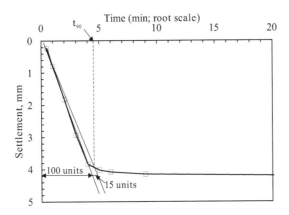

Figure 10.18 Time-settlement data plotted using the root time method.

From Figure 10.18, we will obtain the time (t_{90}) necessary to achieve 90% consolidation (U=90%):

$\sqrt{t_{90}} \approx 4.5\,\text{min} \rightarrow t_{90} \approx 20.25\,\text{min}$

The drainage is provided on both sides, i.e., $H_{dr} = \dfrac{20}{2} = 10\,\text{mm}$

The coefficient of consolidation, $c_v = \dfrac{T_{90} \cdot H_{dr}^2}{t_{90}} = \dfrac{0.848 \cdot 0.01^2}{20.25 \cdot 60} = 6.97 \cdot 10^{-8}\,\text{m}^2\,\text{s}^{-1}$

c2) For $U = 70\% \rightarrow T_{70} = 0.403$ (Table 10.3)

For the clay layer, the drainage exists on both sides (gravel and sand are permeable material), i.e.,

$H_{dr} = 5/2 = 2.5\text{m}$

The time to achieve 70% consolidation equals

$t_{70} = \dfrac{T_{70} \cdot H_{dr}^2}{c_v} = \dfrac{0.403 \cdot 2.5^2}{6.97 \cdot 10^{-8}} \approx 36088222\,\text{s} \approx 13.7\text{months}$

c3) From Figure 10.16, we obtain

$$\sigma_1 = 50kN/m^2 \rightarrow e_1 = 1.76$$

$$\sigma_2 = 100kN/m^2 \rightarrow e_2 = 1.58$$

The coefficient of volume compressibility,

$$m_v = \frac{\Delta e/(1+e_1)}{\sigma_1' - \sigma_2'} = \frac{(1.76-1.58)/(1+1.76)}{100-50} \approx 0.0013\,m^2\,kN^{-1}$$

The coefficient of permeability,

$$k = c_v \cdot m_v \cdot \gamma_w = 6.97 \cdot 10^{-8} \cdot 0.0013 \cdot 9.81 \approx 8.9 \cdot 10^{-10}\,m\,s^{-1}$$

Problem 10.2 The soil profile at a site consists of a 3 m-thick sand layer underlain by a 6 m-thick clay layer, which is underlain by an impermeable rock layer as shown in Figure 10.19. The ground water level is 1 m below the surface.

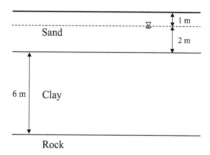

Figure 10.19 Soil profile for Problem 10.2.

Sand layer: the unit weight of sand above the ground water level is 16 kN m⁻³, the unit weight of sand below the ground water level is 18 kN m⁻³.

Clay layer: $Y = 19.5$ kN m⁻³, $m_v = 0.31$ MPa⁻¹, $c_v = 2.6$ m² year⁻¹.

A 3 m compacted fill with the unit weight (γ) of 20 kN m⁻³ will be placed at the ground level.

a) What would be the final consolidation settlement (in *mm*) in the clay layer?
b) How long will it take for 70 mm of consolidation settlement (in *months*)?
c) What would be the consolidation settlement (in *mm*) in one year?

Solutions

a) The settlement of the clay layer under the additional embankment load ($\Delta\sigma' = 3 \cdot 20 = 60$ kN m⁻²) can be estimated as follows:

$$S_c = m_v \cdot \sigma' \cdot H = \frac{0.31}{1000} \cdot 60 \cdot 6000 \approx 111.6\,mm$$

b) If the total settlement is 111.6 mm then the degree of consolidation (U) when the settlement is only 70 mm can be estimated as

$$U = \frac{70}{111.6} \approx 0.627 \text{ or } 62.7\%$$

Resulting in

$T \approx 0.314$ (from Table 10.3)

The drainage of water from the clay layer can only occur through the sand layer above as the rock underneath the clay mass is impermeable, i.e.,

$H_{dr} = 6$ m

The time to reach 70 mm of consolidation settlements will be,

$$t = \frac{T_v \cdot H_{dr}^2}{c_v} = \frac{0.314 \cdot 6^2}{2.6} \approx 4.35 \text{ years or } 52.2 \text{ months}$$

c) We know that t=1 year,

The time factor for t=1 year equals, $T = \dfrac{c_v \cdot t}{H_{dr}^2} = \dfrac{2.6 \cdot 1}{6^2} \approx 0.072$

Giving, $U \approx 30.4\%$ (Table 10.3)
Therefore, the settlement after 1 year can be estimated as

$$S_{1 year} = S_c \cdot U = 111.6 \cdot 0.304 \approx 33.9 \text{mm}$$

10.9 Review quiz

1. Consolidation is a phenomenon that typically occurs in?

    a) rocks
    b) fine-grained soils
    c) coarse-grained soils
    d) all soils

2. If two layers of the same soil have the same degree of consolidation, then the time factors are?

    a) equal
    b) not related
    c) linearly related
    d) none of a-c is correct

3. A 10 m thick soil can only drain from its top surface, the drainage path length (H_{dr} in *m*) is

    a) 10   b) 5   c) 2.5   c) 7.5

4. Which of the following is the compression index?

   a) C_c b) m_v c) c_v d) c_i

5. A layer of clay (thickness of 10 m) is expected to have a total settlement of 1 m. What was the degree of consolidation (in %) when the settlement was 10 cm?

   a) 1   b) 10   c) 99   d) 90

6. A vertical stress of 100 kN m^{-2} is applied to saturated clay, the excess pore water pressure (kN m^{-2}) when time approaches infinity is

   a) 0   b) 50   c) 100   d) infinity

7. What statement is NOT correct?

   a) Consolidation is time dependent
   b) Consolidation occurs due to the expulsion of air
   c) Compression index of clay is generally greater than the one of sand
   d) Consolidation typically occurs in fine-grained soils

8. The log time method to determine the coefficient of consolidation is designed to obtain the time necessary to achieve

   a) 50% consolidation
   b) 75% consolidation
   c) 90% consolidation
   d) 100% consolidation

9. The current vertical effective stress acting on soil is 100 kN m^{-2}. The past maximum vertical effective stress was 200 kN m^{-2}. The overconsolidation ratio is

   a) 0.5   b) 1   c) 2   d) 0

Answers: 1) b   2) a   3) a   4) a   5) b   6) a   7) b   8) a   9) c

Chapter 11

Shear strength of soil

Project relevance: Shear strength of soil is used in slope stability analyses and engineering designs. Before the sand embankment is built at the project site to consolidate the soft alluvial clay, it is necessary to estimate its stability. This chapter will introduce shear strength of soil and factors affecting it. In the end, we will apply this knowledge to compute the factor of safety of the proposed embankment.

11.1 Shear strength of soil in practice

As soil is a weak material, it has a relatively low shear strength compared to rocks, and for this reason, it tends to fail in shear under loads. Natural slopes and man-made embankments may experience stability issues when the shear stress along the failure plane (which is also known as the failure plane or failure surface) exceeds the shear strength of soil (Fig. 11.1). As a result, slope failures such as landslides can occur causing significant economic loss to local communities.

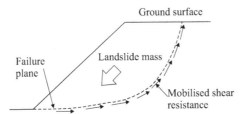

Figure 11.1 A schematic illustration of the failure plane (or *failure surface*) formed in slope and the mobilized shear strength (resistance) of soil acting along the failure plane.

11.2 Shear strength of soil in the laboratory

To determine the shear strength of soil, a series of shear box tests are performed in the laboratory (Fig. 11.2). During such tests, the normal force is vertically applied to the soil specimen (Fig. 11.3) and kept constant throughout the test.

The shear force is then applied in steps and the corresponding values of shear displacement are recorded and plotted as shown in Figure 11.4.

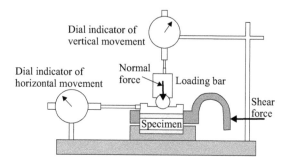

Figure 11.2 Shear box test setup.

Figure 11.3 Sand specimen in the shear box test.

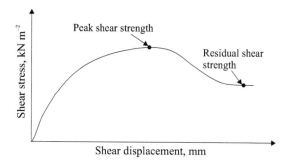

Figure 11.4 Typical shear box test results used to obtain the peak and residual shear strength of soil.

The peak (maximum) and residual shear stresses in each test are obtained (Fig. 11.4) and analyzed in terms of the maximum shear stress vs. the normal stress (Fig. 11.5).

Figure 11.5 presents the results of four shear box tests. The line, which connects the experimental points, is called the strength (or failure) envelope; it indicates the maximum strength of soil at failure (τ_f), and it is mathematically described as shown in Equation 11.1.

$$\tau_f = c + \sigma \tan \phi \tag{11.1}$$

where c is the cohesion and ϕ is the friction angle of soil.

This mathematical expression is known as the Mohr-Coulomb failure criterion; it gives the maximum shear stress (τ_f), which soil can withstand without failure.

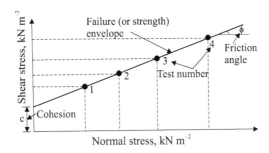

Figure 11.5 Analysis of data from four shear box tests to obtain the strength/failure envelope of soil.

It is commonly assumed that for coarse-grained soils, the cohesion (c) is zero. This type of soil is referred to as "cohesionless" and Equation 11.1 becomes

$$\tau_f = \sigma \tan \phi \tag{11.2}$$

where ϕ is the friction angle which can be determined as follows:

$$\phi = \tan^{-1} \left(\frac{\tau_f}{\sigma} \right) \tag{11.3}$$

Fine-grained soils which typically have cohesion are referred to as *cohesive* soils. Some values of cohesion and friction angle of residual soils originated from hard bedrocks in Queensland (Australia) are given in Table 11.1.

Table 11.1 Properties of residual soils from Queensland (Australia). The values show typical ranges of soil unit weight and strength (after Priddle et al., 2013).

Material property	Parent rock type		
	Mudstone	Sandstone	Tuff
Bulk density, g cm^{-3}	1.9–2.25	1.89–2.25	1.84–2.23
Cohesion, kN m^{-2}	62–360	21–269	24–176
Friction angle, °	2.3–25	4–33	3.2–28.9

11.3 Triaxial compression tests

11.3.1 Deviator stress and confining pressure

In triaxial tests, the soil specimen is first consolidated under an equal all-around pressure of σ_3 (Fig. 11.6). Note that σ_3 is also referred to as the *cell pressure, confining pressure* or *chamber pressure*. The axial stress (σ_1) is then applied until the specimen fails. The deviator stress ("q" or "$\Delta\sigma_d$"), which equals the difference between the principal stresses (σ_1-σ_3), is commonly used to describe the stress conditions at failure.

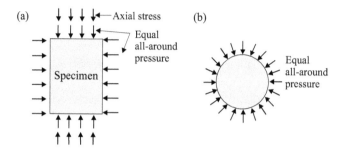

Figure 11.6 Stresses in a triaxial compression test: a) all-around pressure (σ_3) used to consolidate the specimen before the application of axial stress (σ_1); b) a plan view of the all-around pressure.

For example, $q_f = (\Delta\sigma_d)_f = 100$ kN m^{-2} means that the principle stress difference ($\sigma_1 - \sigma_3$) at failure is 100 kN m^{-2}. The term "back pressure" is related to the excess pore water pressure that can be applied to the specimen during testing.

11.3.2 Analysis of triaxial compression tests

Analysis of experimental data from triaxial tests involves plotting the Mohr circle for stress conditions at failure. To determine the shear strength parameters (c and ϕ), the failure/strength envelope, which is a *tangent line* to the circles, is drawn as shown in Figure 11.7. The point of intersection between the failure envelope and the τ – axis gives the value of cohesion (c) while the failure envelope inclination defines the friction angle (ϕ) of the soil.

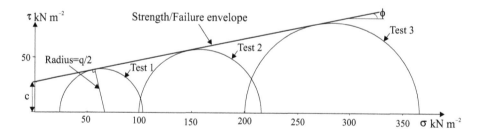

Figure 11.7 Results from three triaxial tests used to determine the strength/failure envelope. The envelope line is tangent to the Mohr circle (forming the right angle with the circle radius). Note that c – cohesion, ϕ – friction angle, q – deviator stress at failure, σ – normal stress, τ – shear stress.

11.3.3 Effective stress conditions

A drained triaxial compression test is performed on the soil specimen in a way where the water drainage is permitted. As the water can freely escape from the specimen during loading, no excess pore water pressure generates ($\Delta u = 0$). In such tests, the effective stress is equal to the total stress ($\sigma_1 = \sigma'_1$, $\sigma_3 = \sigma'_3$) and the load is transferred to the specimen through the soil particles.

Drained conditions are preferable in engineering practice as they provide greater soil strength and result in stability of engineering structures. In the field, drained conditions can be achieved by providing proper drainage. Equation 11.4 gives the relationship between the principal stresses and the shear strength parameters.

$$\sigma'_1 = \sigma'_3 \cdot \tan^2\left(45° + \frac{\phi'}{2}\right) + 2 \cdot c' \cdot \tan\left(45° + \frac{\phi'}{2}\right) \tag{11.4}$$

11.3.4 Undrained triaxial compression tests

In undrained triaxial compression tests, the water drainage is not permitted and the pore water *cannot* escape from the specimen. This creates the undrained stress conditions, resulting in the generation of excess pore water pressure ($\Delta u > 0$). Undrained conditions can occur in soil mass that has a poorly designed drainage system. For low-permeability saturated clayey soils, undrained conditions generally occur immediately after the load application. For natural slopes, heavy rainfalls can lead to undrained conditions as well, especially for low permeability clayey soils.

Figure 11.8 shows the results of two undrained triaxial tests where the Mohr circles are drawn for both total and effective stress conditions. Undrained stress conditions should be avoided in engineering practice, as the excess pore water pressure reduces the effective stress, undermining the strength of soil. Note that the friction angle of soil (ϕ) under total stress conditions is smaller than the one under effective stress conditions (ϕ').

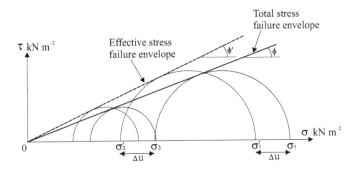

Figure 11.8 Results from two undrained triaxial compression tests analyzed using the total (σ_1 and σ_3) and effective stress conditions (σ_1' and σ_3'). Note that Δu – excess pore water pressure, σ – normal stress, τ – shear stress.

Question: *What type of triaxial tests, drained or undrained, is more useful?*
Answer: The undrained test has the advantage of providing the strength characteristics of soil under both effective (σ') and total (σ) stress conditions while the drained test does not produce any data about the excess pore water generation.

11.4 Stress path concept

The aforementioned analysis using the Mohr circle was performed only for the stress conditions at failure. However, it is also important to observe changes in shear stress from the

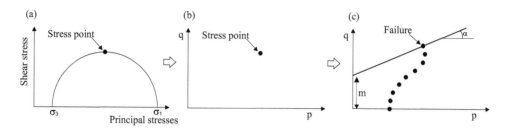

Figure 11.9 Stress path concept: a) the top point of the Mohr circle; b) the stress point in the p vs. q space; c) the stress path reaching failure. The parameters p and q are defined in Equations 11.5 and 11.6.

beginning of load application until the sample fails. To do this, we can employ the stress path concept which is schematically explained in Figure 11.9.

Instead of drawing numerous Mohr circles for each stress increment, we will use only one point that can represent the stress conditions of the Mohr circle. This "stress point" is located on the top of the Mohr circle (Fig. 11.9a) and it is described using the p and q parameters (Fig. 11.9b) which are defined in Equations 11.5 and 11.6.

$$p = \frac{\sigma_1 + \sigma_3}{2} \tag{11.5}$$

$$q = \frac{\sigma_1 - \sigma_3}{2} \tag{11.6}$$

By using the stress path approach, the values of α and m at failure can be obtained as shown in Figure 11.9c. Note that α and m are not soil friction angle (ϕ) and cohesion (c), respectively (Fig. 11.10). However, through the relationship given in Equations 11.7 and 11.8, the friction angle and cohesion can be found.

$$m = c' \cdot \cos \phi' \tag{11.7}$$

$$\phi' = \sin^{-1}(\tan \alpha) \tag{11.8}$$

Note that the stress path is commonly used for effective stress conditions, i.e., p' and q'.

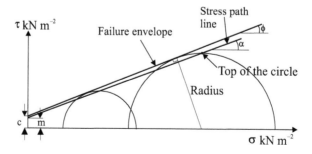

Figure 11.10 The difference between the stress path and Mohr-Coulomb failure criterion.

11.5 Project analysis: shear strength

As the alluvial clay is very soft and saturated, the placement of a 3-m sand fill may cause some stability issues. To estimate the stability of the sand embankment, samples of sand (Sample 1) and alluvial clay (Sample 2) were collected for testing (Fig. 11.11). A series of shear box and triaxial compression tests were performed on these soils and the obtained data are given in Tables 11.2, 11.3 and 11.5. We will analyze the laboratory data and determine the strength characteristics of both samples.

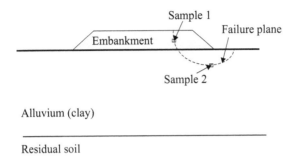

Figure 11.11 Potential failure plane and the location of sample collection.

11.5.1 Laboratory work: shear box tests

Table 11.2 Shear box test details on Sample 1.

Material: Sand	
Bulk density: 1.72 g cm⁻³	Moisture content: 17.2%
Soil particle density: 2.52 g cm⁻³	Sample area: 20 cm²

Table 11.3 Data from a series of shear box tests on Sample 1.

Test N.	Normal Force, kN	Shear force at failure, kN
1	0.15	0.08
2	0.28	0.16
3	0.75	0.40
4	1.20	0.66

Solution

We will convert the normal and shear forces to stresses and record them in Table 11.4. For example, for Test 1, the normal stress is defined as the ratio between the normal force and the specimen area:

$$\sigma = \frac{Force}{Area} = \frac{0.15}{0.0020} = 75 \text{ kN m}^{-2}$$

The shear stress is defined as the ratio of shear force and the specimen area:

$$\text{Shear stress, } \tau = \frac{Force}{Area} = \frac{0.08}{0.0020} = 40\text{kN m}^{-2}$$

Table 11.4 Analysis of data from shear box tests on Sample 1.

Test N.	Normal Force, kN	Normal stress, kN m^{-2}	Shear force at failure, kN	Shear stress, kN m^{-2}
1	0.15	75	0.08	40
2	0.28	140	0.16	80
3	0.75	375	0.40	200
4	1.20	600	0.66	330

Using the normal and shear stresses from Table 11.4, we will plot the laboratory data (Fig. 11.12) and draw the failure envelope to determine the strength characteristics of soil. As the tested soil is sand, its cohesion will be zero.

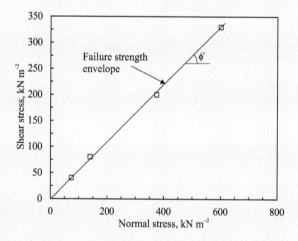

Figure 11.12 Analysis of shear box tests on sand (Sample 1). The failure envelope gives the friction angle (ϕ') of 28.6° and soil cohesion (c) of zero.

Question: The obtained friction angle of this sand is 28.6°, is it high or low?
Answer: The friction angle of sand generally varies from 30–35°, depending on soil mineral composition (quartz or silica), shape of sand grains (round or angular) and specimen density (loose or dense). It can reach up to 45° for angular dense sand.

11.5.2 Laboratory work: triaxial compression tests

Results of three drained triaxial compression tests on soil specimens of Sample 2 are summarized in Table 11.5 in terms of the confining pressure and deviator stress at failure.

Table 11.5 Results of triaxial tests on Sample 2.

Test N	Confining Pressure σ_3, kN m^{-2}	Deviator stress at failure q_f, kN m^{-2}
1	50	102
2	100	131
3	200	162

We will plot the Mohr circle for each test, draw the failure envelope and determine the values of cohesion and friction angle (Fig. 11.13). From the drawing, the cohesion is about 36.6 kN m^{-2} and the friction angle is 9.4°.

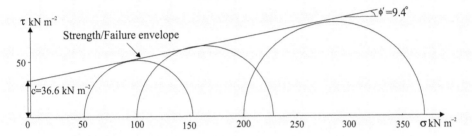

Figure 11.13 Analysis of triaxial tests on Sample 2 using the Mohr circles. The friction angle is 9.4° and the cohesion is 36.6 kN m^{-2}.

Undrained triaxial tests. A saturated specimen (Sample 2) was consolidated in the triaxial cell under a cell pressure of 100 kN m^{-2} (drained conditions). The drainage valve was then closed and a deviator stress was gradually increased from 0 to 112 kN m^{-2} under undrained conditions when failure occurred. Estimate the value of pore water pressure at failure (u_f)?

The stress conditions at failure are schematically shown in Figure 11.14. To solve this problem, we will draw the Mohr circles using the obtained results for the total

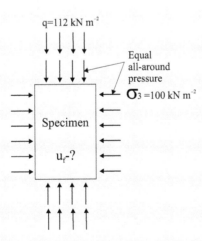

Figure 11.14 Schematic illustration of stress conditions at failure. The deviator stress (q) is 112 kN m^{-2} and the all-around confining pressure (σ_3) is 100 kN m^{-2}.

and effective stress conditions (see Figure 11.13). For the totals stress conditions, the minor principal stress is $\sigma_3 = 100$ kN m^{-2}. The major principal stress equals

$$\sigma_1 = \sigma_3 + q = 100 + 112 = 212 \text{kN m}^{-2}$$

We do not know the principal stresses at failure for the effective stress conditions but we will schematically show them in Figure 11.15 (dotted line). The failure envelope for the effective stress conditions was already obtained from Figure 11.14 using the lab data from Table 11.5. Considering the triangle ABE in Figure 11.15, we can write the following relationship:

$$\sin 9.4^\circ = \frac{BE}{AE} = \frac{56}{c'/\tan 9.4^\circ + (100 - u_f) + 56}$$

where $c' = 36.6$ kN m^{-2} (from Figure 11.13). From this equation, we get

$$u_f \approx 34.2 \text{kN m}^{-2}$$

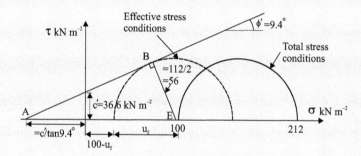

Figure 11.15 Analysis of undrained triaxial tests using the Mohr circles. The failure envelope for the effective stress conditions ($\phi' = 9.4^\circ$ and $c' = 36.6$ kN m^{-2}) was previously found from the drained triaxial tests (see Figure 11.13).

11.5.3 Stress path analysis

We can also solve this problem by using the stress path approach (p' and q'). The values of p' and q' at failure for each test are calculated using Equations 11.5 and 11.6 and given in Table 11.6. We will plot the stress path data of this test in Figure 11.16.

Table 11.6 Values of p' and q'.

σ_3 kN m^{-2}	Deviator stress, kN m^{-2}	σ_1 kN m^{-2}	p', kN m^{-2}	q', kN m^{-2}
50	102	152	101	51
100	131	231	165.5	65.5
200	162	362	281	81

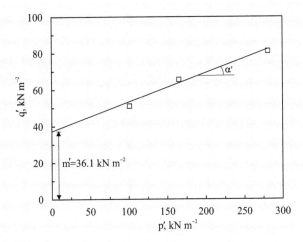

Figure 11.16 Analysis of triaxial tests using the stress path approach.

From Figure 11.16, we obtain

$m' \approx 36.1$ kN m^{-2} and $\tan \alpha' \approx 0.163$

Using Equation 11.8, we get

$\sin \phi' = 0.163 \rightarrow \phi' \approx 9.4°$

From Equation 11.7, we obtain

$$c' = \frac{m'}{\cos \phi'} = \frac{36.1}{\cos 9.4°} = 36.6 \text{ kN m}^{-2}$$

11.6 Problems for practice

Problem 11.1 To perform slope stability analysis of a natural slope (Fig. 11.17), two samples were collected. Sample 1 was collected from the Soil 1 layer and Sample 2 from the Soil 2 layer. The unit weight of Soil 1 was 16 kN m^{-3} and the unit weight of Soil 2 was 18 kN m^{-3}.

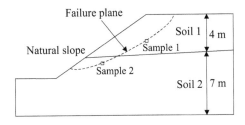

Figure 11.17 Failure plane in natural slope.

A series of direct shear box tests were performed on Sample 1. The area of each sample was 16 cm². The results are summarized in Table 11.7.

a) Plot the obtained results and determine the shear strength parameters of this soil (Soil 1)
b) What would be the maximum shear strength (τ, in $kN\ m^{-2}$) of this soil at a depth of 2 m?

Table 11.7 Data from a series of shear box tests.

Test N.	Normal Force, kN	Peak shear force, kN	Residual shear force, kN
1	0.15	0.12	0.05
2	0.28	0.18	0.1
3	0.75	0.42	0.25
4	1.20	0.60	0.41

Solution

a) Let's determine the stresses at failure for Test 1. Normal stress equals

$$\sigma = \frac{F}{A} = \frac{0.15}{0.0016} \approx 93.75\,\text{kN m}^{-2}$$

Shear stress equals

$$\tau = \frac{F}{A} = \frac{0.12}{0.0016} \approx 75\,\text{kN m}^{-2}$$

Stresses for all four tests are summarized in Table 11.8 and plotted in Figure 11.18

Table 11.8 Analysis of data from shear box tests.

Test N.	Normal Force, kN	Peak shear force, kN	Residual shear force, kN	Normal stress, kN m⁻²	Peak shear stress, kN m⁻²	Residual shear stress, kN m⁻²
1	0.15	0.12	0.05	93.75	75.0	31.25
2	0.28	0.18	0.1	175.00	112.5	62.50
3	0.75	0.42	0.25	468.75	262.5	156.25
4	1.20	0.60	0.41	750.00	375.0	256.25

From the plot in Figure 11.18, we obtain the peak friction angle, $\phi' \approx 24.8°$ and soil cohesion, $c' \approx 34$ kN m⁻². The residual shear strength can be described as $\phi'_r \approx 18.7°$ and $c' = 0$.

Question: *Is the residual strength of soil also important or can it be neglected?*
Answer: Studies show that residual strength is as important especially when engineers deal with slow-moving or reactivated landslides. It is commonly assumed that the soil that reached its residual strength has only its friction component (i.e., friction angle) while the cohesion becomes zero (Skempton, 1985). The residual strength depends on soil mineralogy and it can be very low ($\phi=10°$) for soils with montmorillonite (Gratchev and Sassa, 2015).

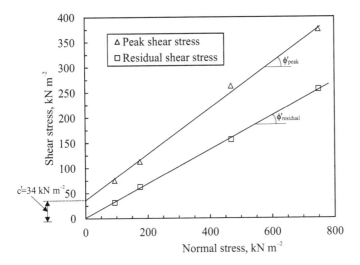

Figure 11.18 Results of four shear box tests plotted as normal stress against shear stress. The failure envelope determines soil friction angle of 24.8° and cohesion of 34 kN m⁻². The residual friction angle is 18.7° and cohesion is zero.

b)   The maximum shear strength of this soil at a depth of 2 m can be assessed as

$$\tau = \sigma \cdot \tan\phi + c$$

where $\sigma = \gamma \cdot H = 16 \cdot 2 = 32\,\text{kN m}^{-2}$

Therefore, the shear strength at 2 m equals

$$\tau = 32 \cdot \tan 24.8 + 34 \approx 49\,\text{kN m}^{-2}$$

Problem 11.2. Consolidated drained triaxial tests on two specimens of Soil 2 produced the following results (Table 11.9). Plot the obtained results using the Mohr circle diagram and determine the values of cohesion (c′) and effective friction angle (ϕ′).

Table 11.9 Results from two triaxial tests on the Soil 2 samples.

Test N.	Confining Pressure σ_3, kN m⁻²	Deviator stress at failure q_f, kN m⁻²
1	50	52
2	200	120

Solution

Results of the triaxial tests from Table 11.9 are plotted in Figure 11.19 to determine the cohesion and friction angle of soil.

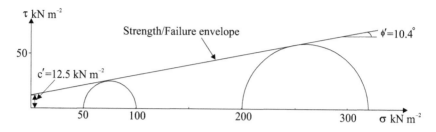

Figure 11.19 Mohr circles and failure envelope for stress conditions at failure: $c' \approx 12.5$ kN m^{-2} and $\phi' \approx 10.4°$.

Another way to solve this problem involves Equation 11.4. We will write the relationship between the principal stresses for each test:

$$102 = 50 \cdot \tan^2\left(45° + \frac{\phi'}{2}\right) + 2 \cdot c' \cdot \tan\left(45° + \frac{\phi'}{2}\right) \tag{a}$$

$$320 = 200 \cdot \tan^2\left(45° + \frac{\phi'}{2}\right) + 2 \cdot c' \cdot \tan\left(45° + \frac{\phi'}{2}\right) \tag{b}$$

As these tests were conducted on the same soil, we can subtract Equation (a) from Equation (b), resulting in

$$218 = 150 \cdot \tan^2\left(45° + \frac{\phi'}{2}\right)$$

Solving this, we get

$$1.45 = \tan^2\left(45° + \frac{\phi'}{2}\right)$$

$$1.2 = \tan\left(45° + \frac{\phi'}{2}\right)$$

$$50.2° = 45° + \frac{\phi'}{2} \rightarrow \phi' \approx 10.4°$$

By substituting ϕ' with 10.4 in Equation (a), we obtain

$$c' \approx 12.5 \text{ kN m}^{-2}$$

Problem 11.3 A specimen of saturated sand was consolidated under all-around pressure of 60 kN m^{-2}. The axial stress was then increased and drainage was prevented. The specimen failed when the axial deviator stress ($\Delta\sigma_d$) reached 50 kN m^{-2}. The pore water pressure at failure was 41.35 kN m^{-2}. Determine:

a) The undrained friction angle (total stress conditions)
b) The drained friction angle (effective stress conditions).

Solution

We know that the tested soil is sand, which means that the soil cohesion equals zero. Let's determine the principal stresses at failure for total stress conditions:

$$\sigma_3 = 60\,\text{kN m}^{-2}$$

$$\sigma_1 = \sigma_1 + (\Delta\sigma_d)_f = 60 + 50 = 110\,\text{kN m}^{-2}$$

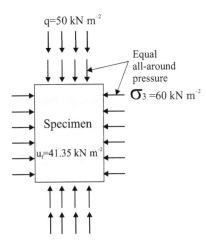

Figure 11.20 Stress condition at failure for the soil specimen from Problem 11.3.

The Mohr circles and failure envelope for total and effective stress conditions are shown in Figure 11.21. From the circle geometry, we get

$$\sin\phi = \frac{(\sigma_1 - \sigma_3)}{(\sigma_1 + \sigma_3)} \approx 0.294$$

The friction angle will be

$$\phi \approx 17.1°$$

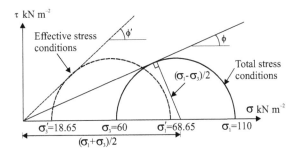

Figure 11.21 Data analysis using the total and effective stress conditions.

Using the effective stress conditions and pore water pressure at failure (u_f), we get

$$\sigma_3' = \sigma_3 - u_f = 60 - 41.35 = 18.65 \text{kN m}^{-2}$$
$$\sigma_1' = \sigma_1 - u_f = 110 - 41.35 = 68.65 \text{ kN m}^{-2}$$

From the Mohr circle drawn for the effective stress conditions (Fig. 11.21), we have

$$\sin \phi = \frac{\left(\sigma_1' - \sigma_3'\right)}{\left(\sigma_1' + \sigma_3'\right)} \approx 0.572$$

Resulting in soil effective friction angle of

$$\phi' \approx 34.9°$$

Problem 11.4 Consolidated drained triaxial tests on three identical specimens of soil produced the following results (Table 11.10).

Table 11.10 Results from triaxial tests.

Test N.	Confining Pressure σ_3, kN m^{-2}	Deviator stress at failure q_f, kN m^{-2}
1	50	52
2	100	82
3	200	120

a) Plot the Mohr circle for each test and determine the values of cohesion and friction angle from your drawing.
b) Plot the obtained results using a stress path approach (p' and q') and determine the values of cohesion and effective friction angle.
c) Another identical specimen of the same soil was consolidated in the triaxial cell under a cell pressure of 100 kN m^{-2} (drained conditions). The drainage valve was then closed and a deviator stress was gradually increased from 0 to 65 kN m^{-2} under *undrained* conditions when failure occurred. Estimate the value of pore water pressure at failure?

Solution

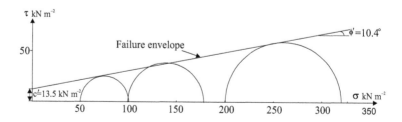

Figure 11.22 Analysis of three triaxial tests to determine soil cohesion (c' = 13.5 kN m^{-2}) and friction angle (ϕ' = 10.4°).

By plotting the Mohr circle for each test as shown in Figure 11.22, we will obtain the soil cohesion and friction angle for effective stress conditions:

$$c' \approx 13.5 \text{kN m}^{-2}, \; \phi' \approx 10.4^{\circ}$$

Table 11.11 Values of p' and q'.

σ_3, kN m^{-2}	Deviator stress, kN m^{-2}	σ_1, kN m^{-2}	p', kN m^{-2}	q', kN m^{-2}
50	52	102	76	26
100	82	182	141	41
200	120	320	260	60

b) From Figure 11.23, we can obtain

$$m' \approx 13.4 \text{kN m}^{-2} \text{ and } \tan \alpha' \approx 0.182$$

The friction angle equals

$$\phi' = \sin^{-1}(\tan \alpha)$$

where $\sin \phi' = 0.182 \rightarrow \phi' \approx 10.5^{\circ}$
Soil cohesion can be obtained using the m parameter as follows:

$$c' = \frac{m'}{\cos \phi'} = \frac{13.4}{\cos 10.5^{\circ}} = 13.7 \text{kN m}^{-2}$$

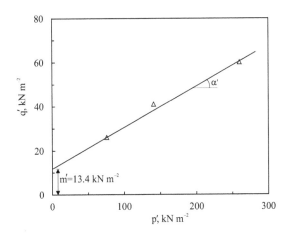

Figure 11.23 Analysis of triaxial tests using the stress path concept to determine the parameters m' and α'.

c) The stress conditions at failure are shown in Figure 11.24. Using the cohesion (c' = 13.7 kN m⁻²) and friction angle (ϕ' = 10.5°) obtained in the previous problem, we will analyze the stresses at failure for effective and total stress conditions (Fig. 11.25).

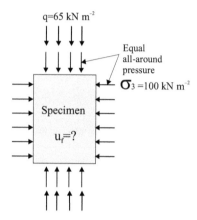

Figure 11.24 Stress conditions at failure.

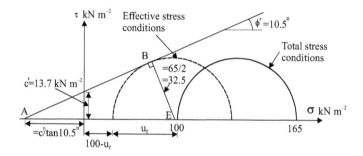

Figure 11.25 Analysis of undrained triaxial test using total and effective stress conditions.

Considering the triangle ABE in Figure 11.25, the following relationship can be obtained:

$$\sin 10.5° = \frac{BE}{AE} = \frac{q/2}{c'/\tan 10.5° + (100 - u_f) + q/2} = \frac{32.5}{73.5 + (100 - u_f) + 32.5}$$

From this equation, we can find

$$u_f \approx 28.1 \text{kN m}^{-2}$$

11.7 Slope stability analysis

Slope stability analysis is a procedure that engineers employ to assess the stability of natural and man-made slopes. It involves the use of factor of safety (FS), which is defined as the ratio between the available and required strength:

$$FS = \frac{Available\ strength}{Required\ strength} \tag{11.9}$$

where the *available strength* is the maximum strength of soil mass as given in Equation (11.1) and the *required strength* is the strength that is necessary to keep the slope stable. When

FS > 1, the slope is considered to be stable; for FS < 1, we assume that the slope fails; if FS = 1, the slope is under *critical conditions*.

Question: When FS = 1, is the slope stable or not?
Answer: The slope is still considered stable. However, we should be cautious with values of FS and remember that it is only a number that may not exactly indicate the slope conditions. Slope stability analysis involves some assumptions and simplifications, which may be different from the existing stress or soil conditions in the field. For this reason, there have been several cases in practice where slopes with FS < 1 were still stable for a long period of time.

11.7.1 Stability of infinite slopes

Many failures occur on slopes where heavily weathered rocks and soil are underlain by hard bedrock (Fig. 11.26). Such landslides are characterized by relatively shallow depths (1–2 m) and triggered by rainfall or earthquakes.

Figure 11.26 Shallow landslide that occurred in heavily weathered mudstone in Niigata Prefecture (Japan). This type of landslide not only disrupts the wellbeing of local communities by blocking major transportation routes, it also causes significant economic damage that involves landslide mass removal and stabilizing of the potentially unstable slopes. More details are given in Gratchev and Towhata (2011).

The stability of such slopes can be analyzed using the infinite slope method as schematically shown in Figure 11.27. Let's consider the following slope and soil characteristics:

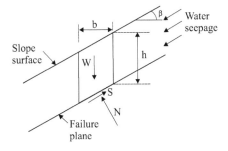

Figure 11.27 Forces acting on infinite slope. W is the weight, N is the normal force, S is the available strength, β is the inclination of slope surface, h is the average thickness of soil layer.

The thickness of the soil layer (h) is 2 m, the slope inclination (β) is 30°. The unit weight (γ) of soil is 20 kN m⁻³, soil cohesion (c) is 10 kN m⁻² and the friction angle (ϕ) is 35°. We will calculate the factor of safety for this slope for two conditions:

a) The slope is dry
b) There is water seepage parallel to the slope

Case A presents the most favorable conditions, as there is no water to decrease the stability of soil mass. The available strength from Equation (11.9) is related to the maximum shear strength of soil mass, which is defined by the Mohr-Coulomb failure criterion (Equation 11.1) while the required strength is related to the gravity force and slope inclination. Considering this, FS can be estimated as follows (Equation 11.10):

$$FS = \frac{c}{\gamma h \sin \beta \cos \beta} + \frac{\tan \phi}{\tan \beta} \qquad (11.10)$$

Using the data on slope geometry and soil strength characteristics, we get

$$FS = \frac{10}{20 \cdot 2 \cdot \sin 30° \cos 30°} + \frac{\tan 35°}{\tan 30°} = 1.79$$

Case B represents the conditions where the presence of water in the soil mass will decrease the available shear strength, resulting in lower values of FS. The stability of slope can be assessed using Equation 11.11:

$$FS = \frac{c}{\gamma h \sin \beta \cos \beta} + \frac{\gamma - \gamma_w}{\gamma} \cdot \frac{\tan \phi}{\tan \beta} \qquad (11.11)$$

Thus, for the slope in Figure 11.27, we will have

$$FS = \frac{10}{20 \cdot 2 \sin 30° \cos 30°} + \frac{20 - 9.81}{20} \cdot \frac{\tan 35°}{\tan 30°} = 1.19$$

For both cases, the factor of safety is greater than one, indicating that the slope is stable. However, in the case of water seepage (Case B), the factor of safety decreased due to the effect of pore water pressure.

Question: *In this analysis, do we know where the failure plane is located?*
Answer: We assume that the failure plane will likely form in the soil mass along the boundary with the relatively hard bedrock.

11.7.2 Limit equilibrium method

Limit equilibrium method (LEM) is a complex analysis in which soil mass is divided into slices and the forces acting on each slice are considered as shown in Figure 11.28.

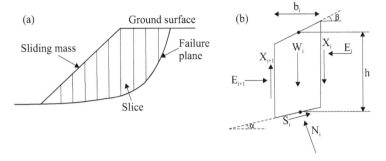

Figure 11.28 Division of potential sliding mass into slices (a); Forces acting on a typical slice (b). W is the weight of slice, N is the normal force, S is the available strength, E and X are the interslice forces acting on the side of each slice, b is the width of slice, α is the inclination of slice base, β is the inclination of slice top, h is the average slice height.

To simplify this complex analysis, a few assumptions are made:

a) Failure occurs along a distinct slip surface
b) The sliding mass moves as an intact body
c) The Mohr-Coulomb failure criterion is satisfied along the slip surface

It is difficult to define the interslice forces, and therefore, more assumptions need to be made to deal with too many unknowns. Different assumptions resulted in several LEM such as the ordinary method of slice (Fellenius method), Bishop simplified method, Janbu method, Morgenstern-Price method and Spencer method (Sharma, 2007).

The Bishop's simplified method appears to be widely used in practice due to its simplicity and satisfactory results. The assumptions used in this method are:

a) The vertical interslice forces are equal and opposite
b) The interslice shear forces are zero

Equation 11.12 gives the mathematical solution of the Bishop's simplified method where the pore water pressure is considered by using the effective shear strength parameters:

$$FS = \frac{Resisting\ moment}{Overturning\ moment} = \frac{\sum_{i=1}^{n}\left[c' \cdot b_i + (W_i - u_i \cdot b_i)\tan\phi'\right]\dfrac{1}{m_{\alpha(i)}}}{\sum_{i=1}^{n}W_i \cdot \sin\alpha_i} \qquad (11.12)$$

where

$$m_{\alpha(i)} = \cos\alpha_i + \frac{\tan\phi' \cdot \sin\alpha_i}{FS} \qquad (11.13)$$

As the term FS is present on both sides of Equation 11.12 we must use a trial-and-error procedure to find the value of FS. A number of failure surfaces must be investigated so that the critical surface that provides the minimum factor of safety can be found. These iterative calculations are time-consuming and can easily take about an hour each. However, using a suitably programmed electronic spreadsheet or commercially available computer programs, the solutions can be determined rapidly.

11.8 Project analysis: slope stability analysis

Let's analyse the stability of the sand embankment built on the soft alluvial clay. We will divide the slope into 10 slices and calculate the slice geometry for each of them. The results are given in Tables 11.12 and 11.13. Note that the value of α in Table 11.12 is positive when the slope of the arc is in the same quadrant as the ground slope.

The shear strength of sand ($\phi = 28.6°$, c = 0) from the embankment was obtained from a series of shear box tests (see Chapter 11.5.1) while the shear strength of alluvial clay was determined from a series of triaxial tests (see Chapter 11.5.2: $\phi = 9.4°$, c = 36.6 kN m^{-2}). For the potential failure plane shown in Figure 11.29, we will calculate the resisting and overturning moments for each slice using Equation 11.12. Table 11.13 shows the final iterated solution that is used to compute the factor of safety.

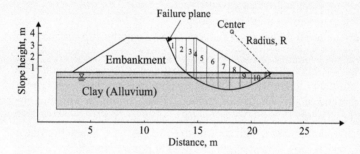

Figure 11.29 Slope stability analysis of the embankment built on the soft alluvial clay. Factor of safety (FS) is 3.48.

Table 11.12 Slice data for the slope conditions from Figure 11.29.

Slice	Width (m)	Height (m)	α (°)	β (°)	Weight (kN)	Cohesion (kN m⁻²)	Friction angle (°)
1	0.97	0.97	−63.3	0	20.03	0	28.6
2	0.97	2.46	−46.8	0	50.75	0	28.6
3	0.89	3.24	−36.1	0	48.23	36.6	9.4
4	0.13	3.54	−31.3	0	9.70	36.6	9.4
5	1.04	3.53	−25.3	30	73.31	36.6	9.4
6	1.04	3.32	−15.0	30	66.47	36.6	9.4
7	1.04	2.91	−5.2	30	56.36	36.6	9.4
8	1.04	2.32	4.4	30	43.20	36.6	9.4
9	1.04	1.56	14.2	30	27.26	36.6	9.4
10	1.26	0.83	25.6	0	16.32	36.6	9.4
11	0.72	0.26	36.1	0	2.95	36.6	9.4

Table 11.13 Solutions using Equation 11.12.

Slice	Resisting force (kN)	Overturning force (kN)
1	18.55	5.32
2	34.63	9.94
3	40.98	11.76
4	7.42	2.13
5	53.85	15.46
6	49.12	14.1
7	46.12	13.24
8	44.33	12.73
9	43.61	12.52
10	55.05	15.8
11	34.61	9.93
Totals	428.3	122.9
	FS=	3.48

Question: *Is this the critical failure plane with the minimum factor of safety?*
Answer: Yes, we ran this analysis using a computer program that analyzed thousands of potential failure planes. The failure plane shown in Figure 11.29 is the critical one with the lowest FS.

Question: *Should we also use the Bishop's simplified method for slope stability analysis?*
Answer: Not necessary, other methods are fine and provide satisfactory solutions as well.

11.9 Review quiz

1. What is the typical range of friction angle for dense sands?

    a) 5–10°   b) 15–20°   c) 25–30°   d) 35–40°

2. Drained triaxial compression test means that the soil specimen is dry.

    a) True   b) False

3. What clay mineral is typically associated with lower shear strength?

    a) kaolinite   b) illite   c) chlorite   d) montmorillonite

4. In general, course-grained soil has no cohesion.

    a) True   b) False

5. The term "back pressure" used for triaxial compression tests means

    a) major principal stress   b) cell pressure
    c) pore water pressure   d) deviator stress at failure

6. In slope stability analysis, the factor of safety (FS) is defined as:

    a) FS = Available strength/Required strength
    b) FS = Required strength/Available strength

7. When FS = 1.5, it means

    a) 100% strength mobilization   b) 67% strength mobilization
    c) 50% strength mobilization   d) 33% strength mobilization

8. The Bishop's simplified slope stability analysis is based on

    a) Limit analysis   b) Limit equilibrium method
    c) Finite element analysis   d) Boundary element analysis

Answers: 1) d   2) b   3) d   4) a   5) c   6) a   7) b   8) b

References

Bergado, D.T., Asakami, H., Alfaro, M.C. & Balasubramaniam, A.S. (1991). Smear effects of vertical drains on soft Bangkok clay. *Journal of Geotechnical Engineering*, 117(10), 1509–1530.

Cogan, J., Gratchev, I.B. & Wang, G. (2018). Rainfall-induced shallow landslides caused by ex-Tropical Cyclone Debbie, 31st March 2017. *Landslides*, 1–7.

Fredlund, D.G. & Rahardjo, H. (1993). *Soil Mechanics for Unsaturated Soils*. John Wiley & Sons.

Gratchev, I.B., Balasubramaniam, A., Oh, E. & Bolton, M. (2012). Experimental study on effectiveness of vertical drains by means of Rowe Cell Apparatus. 447–452. doi:10.3850/978-981-07-3559-3_01-0116

Gratchev, I.B., Irsyam, M., Towhata, I., Muin, B. & Nawir, H. (2011). Geotechnical aspects of the Sumatra earthquake of September 30, 2009, Indonesia. *Soils and foundations*, 51(2), 333–341.

Gratchev, I.B. & Jeng, D. S. (2018). Introducing a project-based assignment in a traditionally taught engineering course. *European Journal of Engineering Education*, 1–12.

Gratchev, I.B., Pitawala, S., Gurung, N. & Errol Monteiro (2018). A chart to estimate CBR of plastic soils. *Australian Geomechanics Journal*, 53(1).

Gratchev, I.B. & Sassa, K. (2015). Shear strength of clay at different shear rates. *Journal of Geotechnical and Geoenvironmental Engineering*, 141(5), 06015002.

Gratchev, I.B., Sassa, K. & Fukuoka, H. (2006). How reliable is the plasticity index for estimating the liquefaction potential of clayey sands? *Journal of Geotechnical and Geoenvironmental Engineering*, 132(1), 124–127.

Gratchev, I.B., Sassa, K., Osipov, V.I. & Sokolov, V.N. (2006). The liquefaction of clayey soils under cyclic loading. *Engineering Geology*, 86(1), 70–84.

Gratchev, I.B., Shokouhi, A. & Balasubramaniam, A. (2014). Feasibility of using fly ash, lime, and bentonite to neutralize acidity of pore fluids. *Environmental Earth Sciences*, 71(8), 3329–3337.

Gratchev, I.B., Surarak, C., Balasubramaniam, A.N. & Oh, E.Y. (2012). Consolidation of soft soil by means of vertical drains: Field and laboratory observations. *11th Australian New Zealand Conference on Geomechanics (ANZ2012)*. pp. 1189–1194.

Gratchev, I.B. & Towhata, I. (2009, January). Effects of acidic contamination on the geotechnical properties of marine soils in Japan. *The Nineteenth International Offshore and Polar Engineering Conference*. International Society of Offshore and Polar Engineers.

Gratchev, I.B. & Towhata, I. (2010). Geotechnical characteristics of volcanic soil from seismically induced Aratozawa landslide, Japan. *Landslides*, 7(4), 503–510.

Gratchev, I.B. & Towhata, I. (2011). Analysis of the mechanisms of slope failures triggered by the 2007 Chuetsu Oki earthquake. *Geotechnical and Geological Engineering*, 29(5), 695.

Gratchev, I.B. & Towhata, I. (2016). Compressibility of soils containing kaolinite in acidic environments. *KSCE Journal of Civil Engineering*, 20(2), 623–630.

Holtz, R.D. & Kovacs, W.D. (1981). *An Introduction to Geotechnical Engineering*. p. 747, ISBN: 0-13-484394-0. https://trid.trb.org/view/214980

Indraratna, B., Balasubramaniam, A.S. & Ratnayake, P. (1994). Performance of embankment stabilized with vertical drains on soft clay. *Journal of Geotechnical Engineering*, 120(2), 257–273.

Indraratna, B. & Redana, I.W. (1998). Laboratory determination of smear zone due to vertical drain installation. *Journal of Geotechnical and Geoenvironmental Engineering*, 124(2), 180–184.

Kim, D.H., Gratchev, I. & Balasubramaniam, A. (2015). A photogrammetric approach for stability analysis of weathered rock slopes. *Geotechnical and Geological Engineering*, 33(3), 443–454.

Look, B.G. (2014). *Handbook of Geotechnical Investigation and Design Tables*. Taylor & Francis Group, London.

Priddle, J., Lacey, D., Look, B., & Gallage, C. (2013). Residual Soil Properties of South East Queensland. *Australian Geomechanics Journal, 48*(1), 67–76.

Sharma, S. (2007). Slope stability assessment using limit equilibrium methods: Landslides and society. In: Turner, A.K., Schuster, R.L. (eds) *Proceedings of the First North American Conference on Landslides*. Omnipress, Madison, WI, USA. pp. 239–260. AEG Special Publication 22. Published by the Association of Environmental and Engineering Geologist, Denver, Colorado 80246.

Skempton, A.W. (1985). Residual strength of clays in landslides, folded strata and the laboratory. *Geotechnique*, 35 (1), 3–18.

Index